"Terms as seemingly straightforward as 'cooperation' and 'effective' can actually be devilishly difficult to define, and harder to tangibly gage, especially in the world of international law and diplomacy. Melissa McCracken takes on the seemingly intractable task of operationalizing both concepts, as applied to the half the land surface of the earth that lies within river basins shared by two or more countries, and does so with aplomb. Invaluable for scholars and practitioners alike, this clear and precise volume will be helpful for anyone working in the realm of water cooperation and diplomacy, and will have useful lessons for those working in other shared resources as well."

Aaron T. Wolf, *Professor of Geography, Oregon State University, USA*

"This book takes the next step in measuring the effectiveness of cooperation over shared water resources. By providing an approach to defining what effective cooperation actually is and to measuring where and how cooperation takes place at what level of effectiveness, it provides great guidance to both researchers and practitioners in ensuring that our shared water resources are being managed equitably and sustainably today and in an increasingly challenging future!"

Susanne Schmeier, *Associate Professor in Water Law and Diplomacy, IHE-Delft, the Netherlands*

"For those who think about and act to improve transboundary water conflicts, 'cooperation' is a most used and abused term. Through her scrutiny of the term, McCracken helps all of us to realise that water 'cooperation' between states can be as instrumental and destructive as it may be positive. Thanks to this book, we can now use the term 'cooperation' accurately, and perhaps move some water arrangements to be more equitable, more sustainable."

Mark Zeitoun, *Professor of Water Security and Policy, University of East Anglia, UK*

Defining Effective Transboundary Water Cooperation

This book establishes a framework for defining transboundary water cooperation and a methodology for evaluating its effectiveness, which will contribute to more effective and therefore successful cooperation processes.

With the increasing focus on transboundary cooperation as a part of the Sustainable Development Goal Framework, there is global recognition of transboundary water cooperation as a tool for improved governance and management of transboundary surface and groundwaters. However, there is not an agreed-upon definition of transboundary water cooperation in the literature or in practice. This book develops the Four Frames of Transboundary Water Cooperation, which is a neutral modular framework for developing context-specific explanatory definitions of transboundary water cooperation in basins and aquifers. The Four Frames of Cooperation are legal, institutional, relational, and outcome. However, we need to move beyond defining cooperation to understand better measures of the quality and effectiveness of cooperative processes. The Weighted Model of Effective Cooperation presents a first step in qualitatively evaluating the effectiveness of transboundary water cooperation. This model defines effective transboundary water cooperation and operationalizes a method to evaluate the effectiveness of cooperative processes over internationally shared waters. Effective cooperation emphasizes the relational and outcome frames of cooperation while working toward equitability and sustainability. Together, the Four Frames of Cooperation and the Weighted Model of Effective Cooperation will improve the understanding of cooperation and encourage a detailed evaluation of the quality, success, and effectiveness of cooperative processes.

This book will be of great interest to students and scholars of water resource management, water governance, and environmental politics. It will also appeal to policymakers and professionals working in the fields of water conflict, water diplomacy, and international cooperation.

Melissa McCracken is an assistant professor of International Environmental Policy at The Fletcher School at Tufts University. Her research focus is on international water policy and management, cooperation and conflict over shared surface and groundwaters, and conflict transformation surrounding environmental resources. Prior to joining The Fletcher School, Melissa was a post-doctoral scholar with the Program in Water Conflict Management and Transformation, as well as the manager of the Transboundary Freshwater Dispute Database housed at Oregon State University.

Earthscan Studies in Water Resource Management

Water, Climate Change and the Boomerang Effect
Unintentional Consequences for Resource Insecurity
Edited by Larry Swatuk and Lars Wirkus

Legal Rights for Rivers
Competition, Collaboration and Water Governance
Erin O'Donnell

Water Allocation Law in New Zealand
Lessons from Australia
Jagdeepkaur Singh-Ladhar

Reciprocity and China's Transboundary Waters
The Law of International Watercourses
David J. Devlaeminck

Water Management in China's Power Sector
Xiawei Liao and Jim W. Hall

Reviving Indigenous Water Management Practices in Morocco
Alternative Pathways to Sustainable Development
Sandrine Simon

Transnational Corporations in Urban Water Governance
Public–Private Partnerships in Mexico and the US
Joyce Valdovinos

Defining Effective Transboundary Water Cooperation
Melissa McCracken

For more information about this series, please visit: www.routledge.com/books/series/ECWRM/

Defining Effective Transboundary Water Cooperation

Melissa McCracken

Routledge
Taylor & Francis Group
LONDON AND NEW YORK

earthscan
from Routledge

First published 2022
by Routledge
2 Park Square, Milton Park, Abingdon, Oxon OX14 4RN

and by Routledge
605 Third Avenue, New York, NY 10158

Routledge is an imprint of the Taylor & Francis Group, an informa business

British Library Cataloguing-in-Publication Data
A catalogue record for this book is available from the British Library

Library of Congress Cataloging-in-Publication Data
Names: McCracken, Melissa, author.
Title: Defining effective transboundary water cooperation / Melissa McCracken.
Description: New York, NY : Routledge, 2022. |
Series: Earthscan studies in water resource management |
Includes bibliographical references and index. |
Identifiers: LCCN 2021045931 (print) | LCCN 2021045932 (ebook) |
ISBN 9780367647803 (hardback) | ISBN 9780367647827 (paperback) |
ISBN 9781003126249 (ebook)
Subjects: LCSH: Water-supply–Management. | Water resources development. |
Water-supply–Political aspects. | Water-supply–Environmental aspects. |
International cooperation.
Classification: LCC HD1691 .M33 2022 (print) | LCC HD1691 (ebook) |
DDC 354.3/66–dc23/eng/20211105
LC record available at https://lccn.loc.gov/2021045931
LC ebook record available at https://lccn.loc.gov/2021045932

ISBN: 978-0-367-64780-3 (hbk)
ISBN: 978-0-367-64782-7 (pbk)
ISBN: 978-1-003-12624-9 (ebk)

DOI: 10.4324/9781003126249

Typeset in Bembo
by Newgen Publishing UK

For my mentors

Contents

Figures

Tables

Acknowledgments

When I first set out on this research project, I was never expecting to write a book. My journey to this point began during my doctoral studies at Oregon State University, where I had the opportunity to intern with the Global Water Partnership. While there, I was given the opportunity to collaborate with Professor Dan Tarlock on a new Technical Background Paper evaluating the methodology of Sustainable Development Goal Indicator 6.5.2. This exploration of the SDG Indicator and other measures of transboundary water cooperation started me down the path toward defining cooperation and effectiveness.

I am grateful for all the support of my mentors and advisors throughout this process and leading up to it, particularly the support and guidance of my doctoral advisor Professor Aaron T. Wolf. I could not have completed this work without the support of my family, friends, and colleagues. Our conversations and discussions on the ideas contained in this book truly helped to deepen, organize, and expand my thinking and ultimately improve the final product.

This book would not be possible without the generosity of the experts I interviewed. I am deeply grateful for their openness, willingness to share their knowledge and experience, as well as their excitement about the topic. This experience reinforces what I have learned about the international water field, which is that it is filled with supportive, engaged, and kind individuals.

I am grateful for the financial support of the Sasakawa Young Leadership Fellowship Fund and the American Association of Geographers Water Resources Specialty Group that enabled me to conduct this research. And lastly to the Global Water Partnership for taking me on as an intern and for allowing the reproduction of text in Chapters 2 and 3 of this book (with revisions and updates) from *Measuring transboundary water cooperation: options for Sustainable Development Goal Target 6.5* (Global Water Partnership, 2017) written by Melissa McCracken for the Global Water Partnership (GWP) for its Technical Committee Background Papers series.

1 Using a common language for transboundary waters

Every day we interact with water, whether through our day-to-day activities, like drinking tea or brushing our teeth, or through our employment, such as in industry or agriculture. For many of us, there is a high likelihood that your water was sourced from a transboundary river basin or aquifer. With 310 transboundary river basins and nearly 600 transboundary aquifers and groundwater bodies intersected by political boundaries (McCracken and Wolf 2019; IGRAC and UNESCO-IHP 2015), internationally shared waters are a vital and ubiquitous component of the world we live in. With the increasing variability and intensity from climate change, rising demand from economic development and population growth, growing threats to water quality and ecosystem health, among other challenges, cooperation over these internationally shared waters is as important as ever. However, cooperation has often been presented as an always positive process for transboundary waters, and there is a lack of agreement over the desired benefits and outcomes of a cooperative process. In short, what makes cooperation effective? By answering these questions, we can improve our understanding of transboundary water cooperation and develop more effective processes to prepare better and adapt to the challenges facing our global waters.

What you are about to read, at its most basic level, is a book about definitions of two essential words in the field of international waters – cooperation and effectiveness. This book presents a new perspective on these definitions and explores these concepts through their complexity, nuance, and context. Academic research is often focused on the generalizability and transferability of results, and to an extent, this book is no different. However, I hope you discover in reading this that definitions of transboundary water cooperation and its effectiveness are more complex and detail-rich than you might expect.

This book sets out to create a framework and a model to define and evaluate effective transboundary water cooperation and is written for students, academics, and practitioners in the field of international waters. Nevertheless, the concepts of cooperation and effectiveness can be translated to other shared natural resources and scales below the international. My aim is that, from this book, you take away a better understanding of the key terms in this field and how to define transboundary water cooperation and effectiveness. Furthermore, it

DOI: 10.4324/9781003126249-1

should allow you to apply these definitions to a cooperative process and better understand the context and nuance that influences and impacts the outcomes and effectiveness of that process. I hope that academics and practitioners find these methods useful to explore and evaluate cooperative processes and that they will be able to use the methods to provide explicit guidance on what actions and policies have promoted effective cooperation over shared waters, as well as what could be done to increase the effectiveness of cooperative processes.

In addition, I have three other points that I hope you take away from this book. First is that you will recognize the importance of place, scale, and context in transboundary water cooperation and that relationships and outcomes matter when developing effective cooperation. Second is you take away that for cooperation to be effective, it must be equitable and sustainable. And third is that you recognize the tendency for cooperation to always be assumed as a positive action. While transboundary water cooperation is vital for sharing water and has the potential to reduce conflict, improve management, and increase resilience in the face of future climate change, not all cooperation is positive. Instead, transboundary water cooperation can be either constructive or destructive, as put forward by the London Water Research Group (Zeitoun et al. 2019). This book aims to shift the narrative of cooperation toward neutral rather than inherently positive. Once cooperation is defined, its effectiveness can be evaluated, and judgment can be made as to whether it is positive or negative.

A tool for navigating the book

As mentioned at the outset, the essence of this book is about definitions. Therefore, it is crucial for our discussion of transboundary water cooperation and effectiveness to have clearly defined terminology used consistently throughout the book. This chapter provides the foundational terminology upon which the exploration of the nuance and complexity of transboundary water cooperation is based, and it serves as a tool for the reader to ensure clarity in the discussion. Please refer back to this chapter as needed while reading this book and reference the definitions.

While at its core, the book is about definitions, in practicality, this book is much broader in its scope, discussion, and ultimate outcomes. As mentioned earlier, this book develops a framework for defining transboundary water cooperation in a specific place, scale, and context, and it establishes a model for evaluating the effectiveness of that cooperation. In order to achieve these goals, the book is divided into three parts, as depicted in Figure 1.1. Part I answers the question: why is a framework for defining transboundary water cooperation and a method for evaluating its effectiveness needed? Chapter 2 identifies the two main challenges regarding transboundary water cooperation. First, there is no consistent definition of transboundary water cooperation, and second, the variety of methods present for measuring cooperation are often attempting to measure its outcomes or effectiveness. However, the number of methods poses a challenge in monitoring progress and making policy recommendations as to

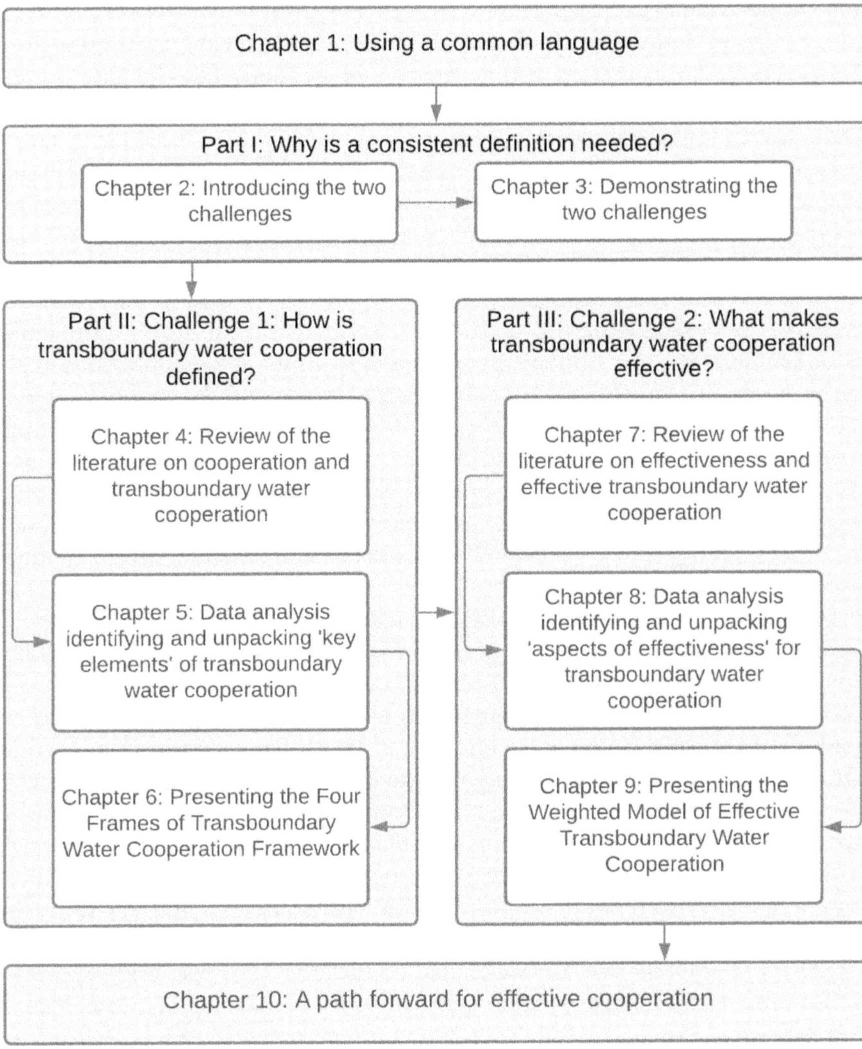

Figure 1.1 Flowchart of the organization and structure of the book.

how to improve the effectiveness of a cooperative process. Chapter 3 exemplifies these two challenges identified in Chapter 2. By comparing three methods for measuring transboundary water cooperation, Chapter 3 demonstrates the complication and confusion that can arise from the use of multiple definitions of transboundary water cooperation and the multiple frameworks for measuring and tracking progress.

Once the two main challenges have been identified and explained, Part II explores the first challenge by answering: How is transboundary water cooperation defined? Chapter 4 begins answering this question by unpacking how cooperation, in a general sense, as well as transboundary water cooperation has been defined in the literature. It presents a review and discussion of how cooperation is used and defined in other fields and the similarities and differences with how transboundary water cooperation is defined in the international waters literature. Using data collected from a qualitative content analysis of literature and interviews with academics and practitioners, Chapter 5 delves into these perspectives on transboundary water cooperation and identifies the key elements that compose it. Using these key elements, Chapter 6 develops a concise definition of transboundary water cooperation, which is listed in the following *Common terminology* section. It then rounds out Part II by recognizing that a statement-style definition of transboundary water cooperation has limitations in that it is difficult to apply to a particular basin or aquifer to determine what is part of the transboundary water cooperation process. The chapter develops a framework for defining transboundary water cooperation by framing the key elements of cooperation into four main categories apparent within definitions of cooperation. Together these four frames represent a means to neutrally define a cooperative process taking into consideration the nuance and variance from a unique context, scale, and place.

The final part of this book, Part III, addresses the second challenge that cooperation does not always establish benefits as expected and that it can contribute to inequity and unsustainability. Therefore, a method is needed to evaluate if a transboundary water cooperative process is effective. Determining what makes transboundary water cooperation effective is the challenge approached by this part. Chapter 7 begins by reviewing the literature on general effectiveness and effective institutional regimes; it then explores how the term 'effective' has been used in the context of transboundary water cooperation in the international waters literature. Similar to Chapter 5, Chapter 8 uses the results of a qualitative content analysis of literature and interviews with academics and practitioners to identify the key aspects of effective transboundary water cooperation, as well as basins and aquifers that have effective cooperative processes and examples of those that lack effective processes. It concludes by developing a statement-style definition of effective transboundary water cooperation, which is listed in the following *Common terminology* section. Part III concludes with Chapter 9, which recognizes that a statement-style definition of effectiveness is not practical for our ultimate goal of evaluating the degree of effectiveness of a cooperative process over shared waters. Therefore, Chapter 9 presents a model and a methodology for applying the model to evaluate effective transboundary water cooperation. The Weighted Model of Effective Cooperation is based on the Four Frames of Transboundary Water Cooperation developed in Chapter 6. The final chapter of this book, Chapter 10, concludes with how the discussions in Parts II and III address the two main challenges identified in Part I. Through this summation, as well as the discussion of the challenges and limitations of the methods developed to define and evaluate effective cooperation, I hope that this chapter – and ultimately this book – allows you to discover that transboundary water cooperation and its effectiveness are more complex and detail-rich than one might expect.

Before diving in, I want to return to the idea that this book is about definitions. International waters, like most fields of study and practice, is heavy with jargon. The language and our common understandings of the terms used imbue value and importance. Our choice of terms has the power to include or exclude and play a role in how groups sharing waters interact. Language can aid in sharing knowledge and increasing efficiency and effectiveness. Therefore, before we begin to discuss how transboundary water cooperation is defined and a model for evaluating its effectiveness, I want to identify some of the key terms used throughout this book and highlight how they are defined. The following section should be used as a reference to gain clarity in the discussion.

Common terminology

A foundational understanding of commonly used terms and concepts is needed. Without a specific and clear understanding of the key terms relevant to the discussion on transboundary water cooperation, such as transboundary basin or arrangement, there is potential for misunderstanding. The commonly used concepts and terms in the field of international waters are defined below:

Watercourse: A watercourse is "a system of surface and groundwaters constituting by virtue of their physical relationship a unitary whole and normally flowing into a common terminus" as defined by the *Convention on the Law of Non-Navigational Uses of International Watercourses* or commonly known as the 1997 UN Watercourses Convention (United Nations 1997). This includes those groundwater bodies that are hydrologically connected and within river basin boundaries. Excluded from the 1997 UN Watercourses Convention definition are the fossil and confined groundwater bodies not hydrologically connected.

Transboundary River Basin: A river basin is the area of land that drains to a common terminus that is an ocean, sea, or terminal inland water body; it is also known as a watershed or catchment. This includes the groundwater bodies that are hydrologically connected to the surface water system. A river basin is transboundary if it contains a perennial tributary that crosses a political boundary between two or more states. (United Nations Centre for Natural Resources, Energy and Transport of the Department of Economic and Social Affairs 1978; Wolf et al. 1999; McCracken and Wolf 2019).

Basin Country Unit (BCU): The area of a transboundary river basin that is within a particular nation. Each transboundary river basin will have at least two BCUs. This area is found when the river basin boundaries are intersected with the political borders (TFDD 2021).

Aquifer. An aquifer is a geologic formation that contains water (International Law Association 2004). More specifically, it is a "permeable water-bearing geological formation underlain by a less permeable layer and the water contained in the saturated zone of the formation" as defined by the United Nations International Law Commission's Draft Articles on the Law of Transboundary Aquifers (United Nations 2016).

Groundwater Body: A groundwater body is defined explicitly by the European Union Water Framework Directive as a "distinct volume of groundwater within an aquifer or aquifers" (European Commission 2000). This distinction between aquifers and groundwater bodies is visible in the categorization in the *2015 Map of the Transboundary Aquifers of the World* (see Figure 2.2). However, in this book, groundwater body is used interchangeably with aquifer.

Groundwater. The water contained within a water-bearing geological formation (International Law Association 2004). In this book, groundwater will be reserved for the water contained within an aquifer. Groundwater can occur in aquifers under two general conditions: confined and unconfined.

Confined: A confined aquifer is overlain by a geologic layer with low hydraulic conductivity; this layer is also known as a confining bed, which restricts the flow of groundwater into or out of the aquifer. An aquifer is confined if water completely fills (full saturation) the pore spaces between the confining beds. It can also be referred to as an artesian aquifer (Heath et al. 2004, 6).

Unconfined: In an unconfined aquifer, water only partly fills the geologic layer. The water table is free to rise and fall, and it is actively involved in the water cycle. These are often called water-table aquifers (Heath et al. 2004, 6).

Transboundary Aquifer or Groundwater Body: A transboundary aquifer (or groundwater body) is intersected by a political boundary and is not hydrologically connected to a surface water system (UNECE 2014). However, in the Draft Articles on the Law of Transboundary Aquifers, an aquifer is transboundary when parts "are situated in different states" (United Nations 2016). Differentiating transboundary aquifers is more complex than river basins, as groundwater bodies can overlap vertically. For those that are vertically overlapping and hydrologically connected, I consider them here as one transboundary aquifer. If they are vertically overlapping and not hydrologically connected – e.g., an unconfined aquifer over a confined aquifer – the groundwater bodies will be considered separate transboundary aquifers unless they are managed conjunctively (UNECE Water Secretariat 2020).

Aquifer Country Unit (ACU): The area of a transboundary aquifer (or groundwater body) that is within a particular state. Given that aquifers have the potential to overlap vertically in the subsurface, the area referred to is the plan view surface area of an aquifer and does not consider the aquifer thickness nor potential volume of water. If overlapping and hydrologically connected, then layers are considered as one ACU. If overlapping and not hydrologically connected, then they are separate ACUs.

Transboundary Area: The total transboundary area within a state is the sum of the surface areas of the BCUs and ACUs in that state. This sum may yield a larger value than the surface area of the country, as the BCUs and ACUs have the potential to overlap. This term is commonly used in reference to the calculation of Sustainable Development Goal Indicator 6.5.2 (see Chapter 3).

Riparian: Relating to or next to a river or stream. It also refers to the nation-states that are within an international river basin and contribute area that drains to the common terminus. For example, the Columbia River Basin is shared between Canada and United States; therefore, Canada is a riparian of the Columbia River Basin.

Hydropolitics: Hydropolitics, like transboundary water cooperation, lacks a single universal definition (Grandi 2020). However, broadly it can be understood as "a systematic study of interstate conflict and cooperation over transboundary water resources" (Elhance 1997, 218).

Institutional Capacity: Institutional capacity includes agreements, treaties, river basin organizations, institutions, and even the positive international relations between riparians (Wolf 2007). Institutional capacity is particularly important in discussions of transboundary water cooperation as it has been shown that institutional capacity can increase the hydropolitical resilience to tensions over shared water (Wolf, Yoffe, and Giordano 2003).

River Basin Organization: A River Basin Organization (RBO) is an "institutionalized form of cooperation that is based on binding international agreements covering the geographically defined area of an international river or lake basins characterized by principles, norms, rules, and governance mechanisms" (Schmeier 2013). RBOs and other institutions on shared waters are central aspects of institutional capacity.

Agreement: An agreement is a bilateral or multilateral formal legal instrument, such as a treaty, amendment, memorandum of understanding, or protocol, between riparian countries regarding transboundary waters.

Arrangement: An arrangement for water cooperation is defined as

> a bilateral or multilateral treaty, convention, agreement or other arrangement, such as a memorandum of understanding, between riparian states that provides a framework for cooperation on transboundary water

> management. Agreements or other kinds of formal arrangements may be interstate, intergovernmental, interministerial, interagency, or between regional authorities.
>
> (UNECE Water Secretariat 2020, 3)

This is a broad definition, but it is centered around the existence of an agreement and is specific to the methodology to calculate the SDG Indicator 6.5.2, which is discussed in Chapter 3.

Cooperation: Cooperation has similar yet deviating definitions depending on the field from which it is derived. In the field of international relations, the most widely cited definition is "when actors adjust their behavior to the actual or anticipated preferences of others through a process of policy coordination" (Milner 1992; Lindblom 1965, 227; Keohane 1984). In the context of this book, I use a definition for general cooperation, which aids in the differentiation between cooperation in general and transboundary water cooperation: Cooperation is a joint action, which can be performed cooperatively and non-cooperatively. Cooperative joint action occurs when participants are willing to complete not only their parts but also assist in additional activities needed for the joint action. Comparatively, non-cooperative joint action is when participants complete their agreed-upon part but will not assist in any additional activities, even those required for the agreed-upon joint action (Tuomela 1993).

Transboundary Water Cooperation: Transboundary water cooperation is the interactions between actors over shared waters that result in establishing mutually beneficial outcomes through a decision-making process; this process could include formal and informal legal and institutional mechanisms depending on the scale and context. This definition is developed by this book (see Chapters 5 and 6).

Effective Transboundary Water Cooperation: Effective transboundary water cooperation achieves equitable results or outcomes through a cooperative decision-making process that has trusting relationships with stakeholder participation and communication. This process is supported by the appropriate legal and institutional arrangements, where the outcomes can be the resolution of the initial problems, reaching the goals of the parties, or complying with mutually agreed-upon principles in a legal arrangement. Effective transboundary water cooperation is equitable and sustainable. This book develops this definition (see Chapters 8 and 9).

Equitable: In the context of this book, equitable is defined in the context of the model developed for evaluating effective transboundary water cooperation – the Weighted Model of Effective Transboundary Water Cooperation. Here, equity or equitable refers to fairness or justice in the cooperative process of sharing transboundary waters. Equitability can be perceived to be achieved when there is an absence of disparities between the actors in the basin in both outcome and process (and perhaps between sectors of use). For each unique basin or aquifer context, equity will be perceived differently. See Chapter 9 for an expanded discussion.

Sustainable: In the context of this book, sustainability is defined as one of the measures – along with equitable – for evaluating whether cooperation is effective. Sustainability is "a condition in which natural and social systems survive and thrive together infinitely (Euston and Gibson 1995)" (Svendsen 2005, 5). However, for this application, I use a definition for sustainable that is aligned with sustainable development. Sustainable and sustainable development are prevalent terms in the literature, as well as within the broader public vocabulary on environmental issues. The most common understanding is balancing trade-offs and desirable goals between the three pillars – social, economic, and environmental – to achieve sustainable development (Purvis, Mao, and Robinson 2019). It is this understanding of sustainable that is applied in this book; see Chapter 9 for an expanded discussion.

My aim in including the following final three terms differs from the previous set of definitions included in this *Common terminology* section. The goal is to provide a common understanding of the key terms used in the field of international waters and, in particular, how they are defined and used throughout this book. However, as many readers will note, the terminology is often used interchangeably or may have overlapping meanings. This is true of transboundary water cooperation – the focus of this book. There are blurred lines between transboundary water cooperation and other terms in the field, namely: water diplomacy, water governance, and water management. Like transboundary water cooperation, these terms also have multiple definitions, and my goal in providing definitions here is not to establish a judgment on what the specific definition should be. Instead, these definitions are a point of comparison, so we may better explore how transboundary water cooperation is defined and determine what it is not. A detailed comparison and discussion of these three terms and their links to transboundary water cooperation are discussed in Chapter 5.

Water Diplomacy: One definition of water diplomacy presented by the Stockholm International Water Institute is that it

> Is a dynamic and evolving process and can support important efforts to achieve peaceful, inclusive and sustainable water cooperation between communities, regions, and countries. It can be tailored and responsive to the uniqueness of each process and geopolitical context.
>
> (SIWI 2016)

In contrast, it can also be defined as

> A theory and practice of implementing adaptive water management for complex water issues. The water diplomacy approach diagnoses water

> problems, identifies intervention points, and proposes sustainable solutions that are sensitive to diverse viewpoints and values, ambiguity and uncertainty as well as changing and competing needs.
>
> ("Water Diplomacy@Tufts" 2019)

From these two definitions, we can see the range in understandings of water diplomacy and the overlap with transboundary water cooperation.

Water Governance: Water governance is not inherently at the transboundary interstate scale; much of the water governance literature focuses on national governance of water at the state, regional, and local levels. One of the most widely cited definitions of water governance is by Rogers and Hall: "Water governance refers to the range of political, social, economic, and administrative systems that are in place to develop and manage water resources, and the delivery of water services, at different levels of society" (Rogers and Hall 2003, 7). However, a growing body of work is shifting to global and interstate management of water. These definitions of global water governance focusing on global norms of water management overlap with the idea of transboundary water cooperation. One definition of global water governance is "the formal and informal institutions, activities, and processes between different actors through which collective interests on water are articulated, differences are mediated, and joint actions are established" (Wiegleb and Bruns 2018).

Water Management: Like water governance, water management does not explicitly focus on internationally shared waters. Water management has evolved from technical and engineered solutions of water problems toward the "activities of analyzing and monitoring, developing and implementing measures to keep the state of water resources within desirable bounds" (Pahl-Wostl et al. 2012, 25). The recognition that many of our shared resources are transboundary has led to expanding the understandings of water management to an explicit internationally shared context. Transboundary water management has been defined as "the application of technical tools to specific water-related questions and challenges to improve water-specific outcomes in watercourses that are transboundary by nature" (Schmeier 2018). In this definition, we can see the roots of water management as a technical and engineering practice, as well as how this would overlap with and require transboundary water cooperation.

One final introductory note: the topic and focus of this book are on *transboundary water cooperation* and *effective transboundary water cooperation*. I use *cooperation* and *effective cooperation* throughout this book, without the 'transboundary water' modifier, to still mean *transboundary water cooperation* or *effective transboundary water cooperation*, respectively. The text explicitly notes when cooperation or effective/effectiveness is being discussed in a broad sense unlinked from transboundary water.

References

Elhance, Arun P. 1997. "Conflict and Cooperation over Water in the Aral Sea Basin." *Studies in Conflict & Terrorism* 20 (2): 207–18. https://doi.org/10.1080/10576109708436034.

European Commission. 2000. *Directive 2000/60/EC of the European Parliament and the Council of October 23, 2000, Establishing a Framework for Community Action in the Field of Water Policy*. Vol. 32000L0060. http://eur-lex.europa.eu/legal-content/EN/TXT/HTML/?uri=CELEX:32000L0060&from=EN.

Euston, Stanley R., and William E. Gibson. 1995. "The Ethic of Sustainability." *Earth Ethics* 5–7.

Grandi, Mattia. 2020. "Hydropolitics." In *Oxford Research Encyclopedia of Environmental Science*, by Mattia Grandi. Oxford University Press. https://doi.org/10.1093/acrefore/9780199389414.013.644.

Heath, Ralph C., North Carolina, Department of Natural Resources and Community Development, and Geological Survey (US). 2004. *Basic Ground-Water Hydrology*. Reston, VA: US Geological Survey.

IGRAC and UNESCO-IHP. 2015. "Transboundary Aquifers of the World Map 2015." International Groundwater Resources Assessment Centre. www.un-igrac.org/resource/transboundary-aquifers-world-map-2015.

International Law Association. 2004. "Berlin Conference on Water Resources Law." International Law Association. http://internationalwaterlaw.org/documents/intldocs/ILA_Berlin_Rules-2004.pdf.

Keohane, Robert O. 1984. *After Hegemony: Cooperation and Discord in the World Political Economy*. Princeton, NJ: Princeton University Press. https://academic.oup.com/ia/article-lookup/doi/10.2307/2617490.

Lindblom, Charles. 1965. *The Intelligence of Democracy*. New York: Free Press.

McCracken, Melissa, and Aaron T. Wolf. 2019. "Updating the Register of International River Basins of the World." *International Journal of Water Resources Development* 35 (5): 732–82. https://doi.org/10.1080/07900627.2019.1572497.

Milner, Helen. 1992. "International Theories of Cooperation among Nations: Strengths and Weaknesses." *World Politics* 44 (3): 466–96. https://doi.org/10.2307/2010546.

Pahl-Wostl, Claudia, Louis Lebel, Christian Knieper, and Elena Nikitina. 2012. "From Applying Panaceas to Mastering Complexity: Toward Adaptive Water Governance in River Basins." *Environmental Science & Policy* 23 (November): 24–34. https://doi.org/10.1016/j.envsci.2012.07.014.

Purvis, Ben, Yong Mao, and Darren Robinson. 2019. "Three Pillars of Sustainability: In Search of Conceptual Origins." *Sustainability Science* 14 (3): 681–95. https://doi.org/10.1007/s11625-018-0627-5.

Rogers, Peter, and Alan W. Hall. 2003. *Effective Water Governance*. Technical Committee Background Papers 7. Stockholm: Global Water Partnership.

Schmeier, Susanne. 2013. *Governing International Watercourses: River Basin Organizations and the Sustainable Governance of Internationally Shared Rivers and Lakes*. New York: Routledge. www.routledge.com/Governing-International-Watercourses-River-Basin-Organizations-and-the/Schmeier/p/book/9780415623582.

———. 2018. "What Is Water Diplomacy and Why Should You Care?" *Global Water Forum* (blog). August 31, 2018. www.globalwaterforum.org/2018/08/31/what-is-water-diplomacy-and-why-should-you-care/.

SIWI. 2016. "International Symposium on Water Diplomacy: An Overview." *Stockholm International Water Institute* (blog). 2016. www.siwi.org/international-symposium-water-diplomacy-overview/.

Svendsen, Mark (Ed). 2005. *Irrigation and River Basin Management: Options for Governance and Institutions*. CABI: International Water Management Institute (IWMI). https://cgspace.cgiar.org/handle/10568/36480.

TFDD. 2021. "Transboundary Freshwater Dispute Database." College of Earth, Ocean and Atmospheric Sciences, Oregon State University. 2021. https://transboundarywaters.science.oregonstate.edu/.

Tuomela, Raimo. 1993. "What Is Cooperation?" *Erkenntnis* 38 (1): 87–101. https://doi.org/10.1007/BF01129023.

UNECE. 2014. "Model Provisions on Transboundary Groundwaters." United Nations Economic Commission for Europe. www.unece.org/fileadmin/DAM/env/water/publications/WAT_model_provisions/ece_mp.wat_40_eng.pdf.

UNECE Water Secretariat. 2020. "Step-by-Step Monitoring Methodology for SDG Indicator 6.5.2 2020." UN-Water. www.unwater.org/publications/step-step-methodology-monitoring-transboundary-cooperation-6-5-2/.

United Nations. 1997. *Convention on the Law of Non-Navigational Uses of International Watercourses. A/RES/51/229*. http://heinonline.org/hol-cgi-bin/get_pdf.cgi?handle=hein.journals/tilj33&section=22.

———. 2016. "The Law of Transboundary Aquifers – Seventy-First Session." Sixth Committee (Legal) - UN General Assembly. 2016. www.un.org/en/ga/sixth/71/transboundary_aquifers.shtml.

United Nations Centre for Natural Resources, Energy and Transport of the Department of Economic and Social Affairs. 1978. *Register of International Rivers*. Vol. 2 (1). Water Supply Management. New York: Published for the United Nations by Pergamon Press.

"Water Diplomacy@Tufts." 2019. http://ase.tufts.edu/igert/waterDiplomacy/documents/keyIdeasPrinciples.pdf.

Wiegleb, Viviana, and Antje Bruns. 2018. "Hydro-Social Arrangements and Paradigmatic Change in Water Governance: An Analysis of the Sustainable Development Goals (SDGs)." *Sustainability Science* 13 (4): 1155–66. https://doi.org/10.1007/s11625-017-0518-1.

Wolf, Aaron T. 2007. "Shared Waters: Conflict and Cooperation." *Annual Review of Environment and Resources* 32 (1): 241–69. https://doi.org/10.1146/annurev.energy.32.041006.101434.

Wolf, Aaron T., Jeffrey A. Natharius, Jeffrey J. Danielson, Brian S. Ward, and Jan K. Pender. 1999. "International River Basins of the World." *International Journal of Water Resources Development* 15 (4): 387–427. https://doi.org/10.1080/07900629948682.

Wolf, Aaron T., Shira B. Yoffe, and Mark Giordano. 2003. "International Waters: Identifying Basins at Risk." *Water Policy* 5 (1): 29–60. https://doi.org/10.2166/wp.2003.0002.

Zeitoun, Mark, Naho Mirumachi, Jeroen Warner, Matthew Kirkegaard, and Ana Cascão. 2019. "Analysis for Water Conflict Transformation." *Water International*, May, 1–20. https://doi.org/10.1080/02508060.2019.1607479.

Part I

Why is a consistent definition to measure transboundary water cooperation needed?

2 Presenting the two challenges

Defining transboundary water cooperation and measuring its effectiveness[1]

The practice of transboundary water cooperation is quite old, with agreements over shared waters dating back to 2500 BCE, most commonly over navigational issues. Cooperation over water for the purposes of water management has increased since 1814 (Hamner and Wolf 1997). In the last several decades, transboundary water cooperation is continued to increase in prominence, such as with the signing of international conventions on transboundary watercourses, the expansion of organizations and institutions focusing on shared waters, and even the inclusion of transboundary water cooperation within the Sustainable Development Goal framework.

At first glance, the promotion of transboundary water cooperation seems like an admirable and worthy goal for states to work toward. Improving cooperation[2] over shared waters has the potential to improve equitable access to water, improve riparian environments, increase global water security, or help solve the 'Water War' crisis.[3] However, establishing and maintaining transboundary water cooperation is a challenging and complex task. This global awareness and emphasis have led to increased efforts to measure and track progress toward achieving transboundary water cooperation. Before accurate measurement can occur, we must ask: What is transboundary water cooperation? The challenge in answering this question lies in the fact that there is no single accepted definition of transboundary water cooperation (Tarlock 2015; Saruchera and Lautze 2015). Without a single accepted definition, multiple conceptualizations of the term are used to devise just as many measures for tracking progress. Having these multiple measures can impact how countries, donors, international organizations, and the greater international system perceive a basin, aquifer, or country's progress toward achieving cooperation. Furthermore, if the goal is to achieve cooperation over transboundary waters, cooperation must be effective – it must have a measurable outcome. However, as with the varying definitions of transboundary water cooperation, what does effectiveness mean, and how is it measured? Part I of this book, including this chapter and Chapter 3, introduces and demonstrates these two main challenges facing the field of international waters, particularly those focused on transboundary water cooperation. Before we dive into exploring the two challenges, this chapter sets

DOI: 10.4324/9781003126249-3

the stage by introducing transboundary waters and the context of international water law that transboundary water cooperation often occurs within.

Transboundary surface and groundwaters

Transboundary waters are surface and groundwaters that cross political, administrative, local, and other types of boundaries. In the context of this book, the boundaries addressed are *international* political boundaries. Wolf et al. (1999) define transboundary river basins as areas that contribute both surface and groundwater to streams that drain to an ocean, sea, or terminal lake – where perennial water intersects a political boundary. There are 310 international transboundary river basins globally (see Figure 2.1), which encompass 47.1% of the world's land surface (McCracken and Wolf 2019) and cross the boundaries of two or more nations. These river systems contribute a significant proportion – approximately 60% – of the freshwater river discharge worldwide (Wolf et al. 1999; Wolf 2007). Most of the world's largest, and often most heavily depended on, rivers cross international borders – these include the Ganges-Brahmaputra-Meghna, Amazon, Indus, Nile, and Colorado rivers. With about 45% of the world's population residing in these basins, equitable sharing and sustainable use are vital for maintaining and increasing the water security for the majority of the world's population (TFDD 2021).

Transboundary waters also include shared groundwater. Defining transboundary groundwaters is more complex than defining transboundary surface waters, given the difficulty in determining the extent of groundwater bodies. Unlike surface waters, the extent of groundwater bodies cannot be determined by physical observation. The above definition of transboundary river basins includes groundwater. These groundwater bodies are shallow, unconfined geologic units that are hydrologically connected to the surface water system, and their extents are contained within the delineation of river basin boundaries. Alternatively, geologic units that are not hydrologically connected to surface water, such as confined and fossil aquifers, can also be transboundary groundwaters. The extent of these groundwater bodies is not related to river basin boundaries.

The definition of transboundary groundwaters is not consistent through legal texts and literature, but the legal distinction between the two types of groundwater bodies is important. The 1997 UN Watercourses Convention (Article 1) defines a watercourse as "a system of surface and groundwaters constituting by virtue of their physical relationship a unitary whole and normally flowing into a common terminus" (United Nations 1997). This definition refers to those groundwater bodies that are hydrologically connected and within river basin boundaries. Excluded from the 1997 UN Watercourses Convention are the fossil and confined groundwater bodies that are not hydrologically connected. These bodies are defined in several variations in legal text and literature. The UNECE Water Convention (Article 2) defines groundwaters as both confined and unconfined aquifers (UNECE 2016b). The 2004 Berlin Rules on Water

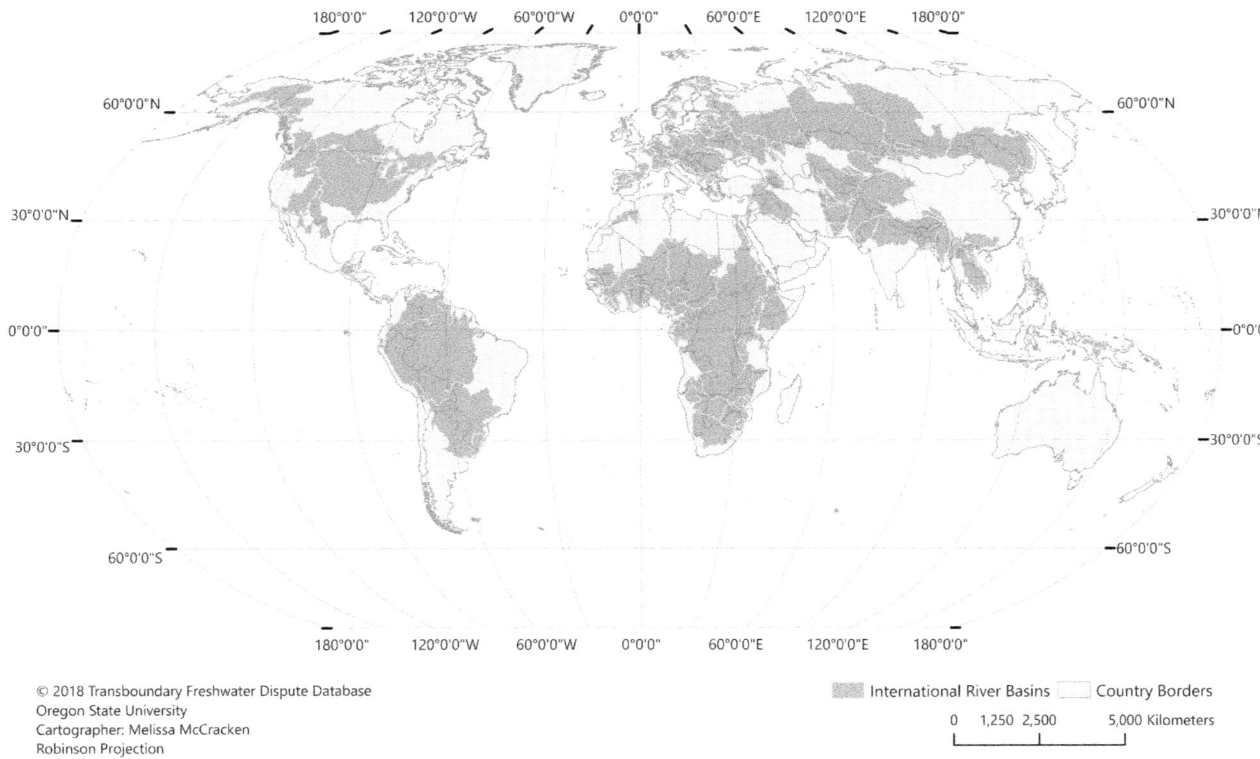

Figure 2.1 International river basins of the world. Product of the Transboundary Freshwater Dispute Database, College of Earth, Ocean, and Atmospheric Sciences, Oregon State University. Additional information about the TFDD can be found at: http://transboundarywaters.science.oregonstate.edu.

Resource Law define an aquifer as a geologic formation that contains water but excludes the water contained within the aquifer from the definition, while the term groundwater is reserved for the water contained within the aquifer (International Law Association 2004). This is similar to the definition used in the draft *Law of Transboundary Aquifers*, where an aquifer is "a permeable water-bearing geological formation underlain by a less permeable layer," but the resolution includes the water contained in the formation within its definition (UN General Assembly 2013). For our purpose here, the definition of an aquifer is consistent with the Berlin Rules; however, the term groundwater body may also be used interchangeably.[4] Transboundary groundwaters or transboundary aquifers can then be defined as aquifers that are intersected by state boundaries (UNECE 2014).

Groundwater has been, and is increasingly becoming, a vital resource for domestic and agricultural water supply. Globally, groundwater represents 99% of all liquid freshwater (UNESCO-IHP and UNEP 2016). According to the latest transboundary aquifer inventory completed by UNESCO IGRAC, there are 592 transboundary aquifers (see Figure 2.2) (IGRAC and UNESCO-IHP 2015). With the difficulty in delineating aquifers, the boundaries are often poorly known, and many are unknown. There may also be further difficulty in determining hydrologic connectivity between aquifers. Therefore, countries may be unaware that groundwater resources are shared, which could increase their vulnerability toward the unsustainable use of these groundwater resources. With the future role that groundwater is likely to play in mitigating and buffering impacts from climate change and rapid human development and population growth, it is important to have an accurate inventory of international aquifers and aquifer systems. Without knowledge of transboundary aquifer systems, transboundary cooperation and management are limited. As of 2021, only seven transboundary aquifers[5] have a formal agreement (TFDD 2021), and a few others have informal agreements. The lack of governance and institutional frameworks on shared groundwater could become a significant impediment to sustainable development (UNESCO-IHP and UNEP 2016).

Rivers and groundwaters traverse landscapes irrespective of political borders. These borders add political complexity to water management with already diverse interests and values. Globally, water demands for domestic use, agriculture, and industry are increasing with population growth, economic development, and changing lifestyles, despite increases in efficiency. Furthermore, of the available supplies, water quality of both surface and groundwaters are at risk of contamination. Much of the global water infrastructure is aging and will need replacement. Due to these factors and the varied distribution of water, water scarcity and water stress are strong arguments for water conflict. Water is often perceived as a source of conflict under these arguments; with a reduction in available water resources, such as from global climate change, there is more competition for access, which leads to a greater potential for conflict. When considering the implications and potential impacts to global water security, the Malthusian arguments warning of increased conflict are compelling. Much of

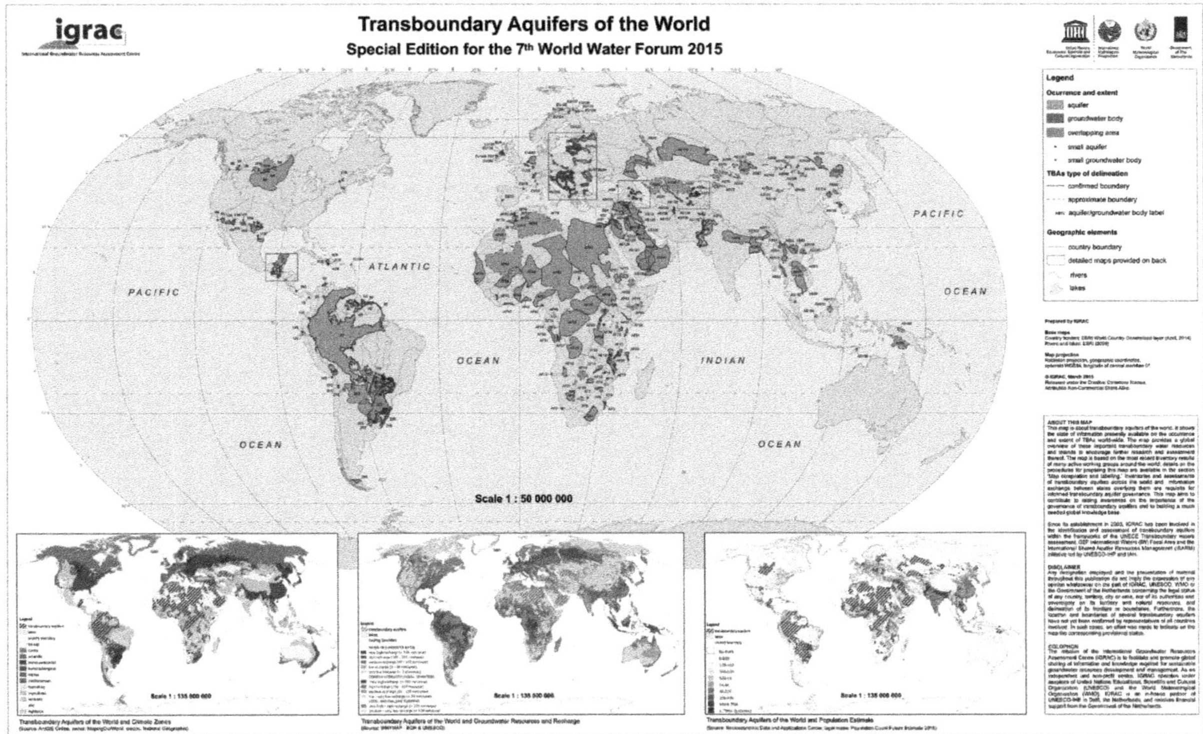

Figure 2.2 Transboundary aquifers of the world.

Source: UNESCO-IHP (UNESCO International Hydrological Programme), 2015. Transboundary Aquifers of the World [map]. Edition 2020. Scale 1:50 000 000. Delft, Netherlands: IGRAC, 2015.

the water conflict literature claims a threat from water scarcity, which would directly lead to warfare between nation-states (Cooley 1984; Committee on Foreign Relations 2011; Yoffe and Wolf 1999; Swain 2015). This water as a source of conflict idea has evolved into an alternative strand of thought. Research and history have shown that while shared water can – and does – lead to tensions and localized hostilities, it does not lead to all-out war between states; rather, the larger the geographic scale, the less likely it is that an intense conflict would occur (Wolf 1998). Tensions, or flashpoints, can encourage disputing states to enter into negotiations which may result in cooperative arrangements (Wolf 2007). Increasing national water security can be obtained through cooperation with co-riparians; sharing water and maintaining water security are not mutually exclusive, even in arid, water-scarce areas. In fact, studies have found that despite the perception of dry areas being more prone to conflict, conflict over water is not any more likely to occur in arid areas than in more humid environments (Wolf, Yoffe, and Giordano 2003). Dinar et al. (2015) take this further and show that high water variability actually lends to more cooperative behavior up to a certain threshold.

Cooperation over shared waters is more prevalent for several reasons, including shared interests and/or the existence of institutional arrangements or particular institutional design features (Wolf 1998; Yoffe and Wolf 1999; Mitchell and Zawahri 2015). As states develop international waters further, they will continue to encounter complexity requiring cooperation and negotiation to prevent loss to the riparians or the resource (Mitchell and Zawahri 2015). Transboundary cooperation presents basin states with opportunities to negotiate solutions that meet shared interests to the benefit of both (or multiple) riparian states (Sadoff and Grey 2002); for example, a collaborative dam project in an upstream country could provide hydropower and irrigation benefits to the downstream country. Shared interests and the benefits of cooperative arrangements are readily visible in past treaties and negotiated agreements between countries (Wolf 1998; Giordano et al. 2013). Once these cooperative arrangements are in place, such as through a treaty, they are found to be resilient even while co-riparians are engaged in a hostile conflict over other issues (Wolf 1998; Yoffe and Wolf 1999; Wolf 2007). Treaties, for example, have reduced the potential for conflict in many regions, but they can come under pressure from political and resource changes in the riparian states (Gleick 1993). Hydropolitics – and the associated hydropolitical resilience or vulnerability – relates to the potential for conflict to occur over shared waters and the ability for institutions to manage the shared resources in a politically sustainable manner. Cooperation and conflict do not occur in isolation from one another, as a basin is not either in a state of conflict or in a state of cooperation (Zeitoun and Mirumachi 2008). Whether a basin is considered hydropolitically vulnerable is linked to the institutional capacity and rate of change within the basin, including inter alia change in physical or institutional systems. The Basins at Risk project found that "the likelihood and intensity of dispute rises as the rate of change within a basin exceeds the institutional capacity to absorb that change" (Wolf, Yoffe, and Giordano 2003, 51; Wolf, Stahl, and Macomber 2003). Institutional capacity, whether treaties, agreements, river basin organizations (RBOs), or positive international relations, is an important parameter for indicating the potential for dispute over shared waters, potentially more so than physical conditions of the basin (Wolf 2007). The ability of the shared water system to be hydropolitically resilient is then dependent on the adequate institutional capacity in place to absorb rapid change within the system; this includes rapid institutional changes, in addition to rapid physical changes. Brochmann (2012) takes this further by showing that the signing of treaties on shared river basins does create an increased potential for future water cooperation. This argument is based on the fact that treaties cover a variety of issues and are intended to improve cooperation generally, in addition to solving specific disputes (Brochmann 2012).

Cooperation in the context of international water law

Cooperation between states is essential. Given this emphasis, having a common understanding of what cooperation is, its origins, and what frameworks exist to guide and influence future cooperation is important if we are to achieve transboundary water cooperation.

Cooperation is coordination between states at a level where they collaborate to achieve a common interest that results in mutual benefits for both states (Zartman 2008 in Leb 2015). This common goal could not be achieved unilaterally by a single state. This definition follows the reasoning discussed above, that there is less conflict that occurs over politically shared waters than expected; states with shared interests and potential for mutually derived benefits are inclined to cooperate, rather than dispute, particularly given the limited gains to be made through a dispute.

With reciprocity and good faith, cooperative behavior between states has developed, leading to interstate cooperation becoming an element of customary international law (Leb 2015). With the establishment of the United Nations, customary interstate cooperation was formalized as a duty to cooperate within the UN Charter. The Charter defines the purpose of the UN "to achieve international cooperation in solving international problems of an economic, social, cultural, or humanitarian character, and in promoting and encouraging respect for human rights and for fundamental freedoms for all without distinction as to race, sex, language, or religion," with specific articles outlining cooperation duties[6] (United Nations 1945, chaps. 1, Article 1(3)). The duty to cooperate is not a hard rule and has limited means to be enforced or encouraged. However, international law has incorporated the duty to cooperate into the norms, principles, and rules established to govern state-to-state relations (Wouters 2013). The UN General Assembly has adopted a resolution that specifically addresses the duty to cooperate as it relates to transboundary water during the creation of the 2013 International Year of Water Cooperation[7] (United Nations 2010).

International law, while not the only mechanism to encourage transboundary water cooperation, does provide a framework to address water sharing and water disputes. Ultimately, the rule of law facilitates cooperation through a framework for governing interstate relations; through a means to integrate across scales, sectors, and disciplines; and through the provision of substantive and procedural rules or legal norms for implementation (Wouters 2013). Substantive rules include those that establish the rights and obligations of states sharing waters, that is, substance. Procedural rules provide the means to implement the substantive rules, that is, procedure. Several global instruments are in place that establish international water law through the duty to cooperate and provide a set of procedural and substantive norms to assist states in governing their interactions over shared waters.

The most important instrument is arguably the *1997 UN Convention on the Law of the Non-Navigational Uses of International Waters* (hereafter UN Watercourses Convention) (United Nations 1997). Adopted in 1997 by the UN General Assembly, the UN Watercourses Convention is considered to be the first legal instrument at the international level to set a duty to cooperate[8] within international water law (Leb 2013). The 35th state (Vietnam) acceded to the convention, bringing it into force in 2014 (United Nations 2016a). The UN Watercourses Convention has several general principles of both substantive and procedural nature, including two substantive cornerstone principles. Through Article 5, states are expected to use shared international waters in a manner that is equitable and reasonable, as well as participate in their use, development, and protection. The second cornerstone is in Article 7, where states shall take measures to prevent significant harm to other riparians when utilizing shared waters (United Nations 1997). The UN Watercourses Convention spells out several other principles: the obligation to regularly exchange data and information (Article 8); the rule that no use has inherent priority over another use in the absence of an agreement (Article 10); the requirement for prior, timely notification of a planned measure (Part III); and the peaceful settlement of a dispute per an agreement or the provision in the convention (Article 33) (United Nations 1997).

In February 2003, the *1992 UNECE Convention on the Protection and Use of Transboundary Watercourses and International Lakes* (hereafter UNECE Water Convention) was amended to allow non-UNECE member countries to accede to the convention.[9] With this opening in March 2016, the UNECE Water Convention is the second international convention to deal with transboundary freshwater (UNECE 2016b). The primary purpose of the UNECE Water Convention is "to strengthen local, national, and regional measures to protect and ensure the ecologically sustainable use of transboundary surface waters and groundwaters" (UNECE 2004, 6). The UNECE Water Convention has a three-pronged structure that is based on these obligations: to prevent, control, and reduce transboundary impact; the requirement for equitable and reasonable use; and the duty to cooperate with co-riparians (UNECE 2016a). These obligations are divided into rules that apply to all parties to the convention (Part I) and requirements that apply to co-riparians sharing the same international waters (Part II) (United Nations Economic Commission for Europe 2013).

The two global conventions are not mutually exclusive and have elements that allow them to complement one another. Generally, the UNECE Water Convention is more detailed and includes a greater degree of procedural rules than the UN Watercourses Convention, which may be considered as giving more detail to substantive principles, for example, 'equitable and reasonable use' and 'no significant harm' (Tanzi 2000; UN Watercourses Convention n.d.). With respect to the duty to cooperate, both conventions express procedural rules as a means for cooperation, for example, data exchange, notification, and joint monitoring. The UNECE Water Convention, however, centers on institutionalized cooperation, as it mandates states enter into bilateral/multilateral agreements and establish joint management bodies (Article 9) (United Nations Economic Commission for Europe 2013); these are not compulsory within the UN Watercourses Convention (Tanzi 2000). Both conventions address groundwater in their definitions of transboundary waters, as mentioned above. However, whether an aquifer that is not hydrologically connected to surface waters is included under the purview of the convention is questionable and generally considered outside its scope (Tanzi 2000).

In 2008,[10] the UN General Assembly adopted the *Draft Articles on the Law of Transboundary Aquifers*; on 4 November 2016, the Draft Articles were again placed on the provisional agenda of a future session for consideration as to whether they should be adopted or whether consideration should be given to transform the Draft Articles into a convention or comparable action[11] (United Nations 2016c). The UN Watercourses Convention was the basis for the Draft Articles, and there are similar provisions between the two instruments. For example, the Draft Articles also include the general principles of equitable and reasonable use, the obligation not to cause significant harm, and the duty to cooperate. However, contrary to the UN Watercourses Convention, the Draft Articles include a general principle (Article 3) of state sovereignty over the portion of a transboundary aquifer within its land (Yamada 2011; Behrmann and Stephan 2010).

In addition to the Draft Articles, the UNECE has issued *Model Provisions on Transboundary Groundwaters*, which provides guidance on the application of the UNECE Water Convention to transboundary groundwaters. The Model Provisions were adopted during the sixth Meeting of the Parties and are recommended for both member and non-member states to use when developing or amending agreements on transboundary groundwaters (UNECE 2014).

In practice, most transboundary water cooperation is governed by treaties or bilateral/multilateral agreements between co-riparians, although few agreements have been signed for transboundary groundwaters. States have historically adopted treaties over shared waters;[12] it is the precedent of cooperation and the establishment of treaties that create the duty to cooperate as a customary international law (Leb 2013). With the global conventions, international water law helps encourage cooperation by providing common rules that govern state-to-state relationships. Along with developing cooperation through trust and reciprocity, treaties create stability and predictability in state-to-state relationships (Leb 2015). Through the substantive principles and procedural duties, international water law encourages cooperation; however, there is not enough stress on cooperation, and unilateral action is often more desirable to states than the mutual benefits gained through cooperation (Tarlock 2015). The degree of cooperation that occurs between states is ultimately determined by the will of the national government. Collaboration can exist without a treaty over shared waters, but it is generally less stable or resilient (Leb 2015; Wolf 1998). Universal frameworks encourage the establishment of an agreement or arrangement to share transboundary water; the UNECE Water Convention even goes so far as to make institutionalized cooperation through an agreement or joint body compulsory. However, there is no international enforcement mechanism to hold states accountable to begin or continue to cooperate per existing agreements.

Customary international law and the universal frameworks within international water law help to establish a more formal and institutional conceptualization of transboundary water cooperation, with a greater emphasis on procedural duties. Transboundary water cooperation – procedural and substantive duties – is reflected in the establishment of agreements over shared surface- and groundwaters, which are found to be indicators of hydropolitical resilience and potential for future cooperation. However, transboundary cooperation can exist beyond treaties and agreements between basin states; political will and non-state actors can contribute to transboundary water cooperation. In fact, many treaties are allowing the participation of non-state actors (Conca, Fengshi Wu, and Cigi Mei 2006). Therefore, measuring the extent of transboundary cooperation is complex, given the consideration as to the type – procedural or substantive – of cooperation and the level of formality – governmental or non-state actors.

Beyond international water law, hydropolitics plays a role in influencing the extent, type, and willingness to cooperate. Hydropolitics relate to the ability of institutions to manage their shared water resources in a politically sustainable manner and how that affects the potential for conflict to occur over shared water (Wolf 2007). It is a component of geopolitical theory, which encapsulates the relationships between politics and space at various scales (Cresswell 2013). Geopolitics and hydropolitics are used as a method to understand how states interact to develop cooperation over shared waters and the role that politics plays in supporting or preventing cooperation. Transboundary water cooperation and hydropolitics influence and are influenced by the water governance of riparian states. Water governance is defined as the "range of political, social, economic and administrative systems that are in place to develop and manage water resources, and the delivery of water services, at different levels of society" (Rogers and Hall 2003). Understanding the role of water governance in transboundary water cooperation will be important for determining its effectiveness, as governance may impact the ability for cooperative efforts to be effective, which leads us to the main challenges this book addresses.

Presenting the two challenges

As discussed above, transboundary cooperation is important, given the global extent of transboundary waters and the population's dependence on them. When including shared groundwaters, this extent and dependence on transboundary waters increase even further. The ubiquity of shared waters underscores the need for transboundary cooperation. While cooperation has historically occurred over transboundary waters and has been shown to increase the potential for benefits from the resource, we still lack a clear understanding of what constitutes transboundary water cooperation (Challenge 1) and what aspects make it the most effective (Challenge 2). Instead, there are multiple understandings, which present complications, including how the definition could be manipulated to support an underlying premise or viewpoint on what constitutes transboundary water cooperation.

Challenge 1: Multiple understandings of transboundary water cooperation

There are many attempts to define and measure transboundary water cooperation. One of the most common measurements of transboundary water cooperation is simply the presence of an agreement, treaty, or formal arrangement over the transboundary basin and/or aquifer. This interpretation stems from research that illustrates that adequate institutional capacity is needed to adapt and respond to changes (Wolf, Yoffe, and Giordano 2003). Treaties, as a form of institutional capacity, create an increase in the potential for future water cooperation (Brochmann 2012). To demonstrate this interpretation of transboundary water cooperation, Table 2.1 shows the extent of the transboundary basin area in Bangladesh, Honduras, and Uganda that has transboundary water cooperation. In this interpretation, cooperation over shared waters within a riparian strictly implies whether an agreement – a bilateral or multilateral formal legal instrument, such as a treaty, amendment, or protocol, between riparian countries regarding transboundary waters – exists for the transboundary river basin or aquifer area.

Using the Transboundary Freshwater Dispute Database (TFDD 2021) for the area of transboundary river basins and the International Groundwater Resources Assessment Centre[13] (IGRAC and UNESCO-IHP 2015) dataset, the area of the transboundary aquifers in each country,[14] Table 2.1 presents the proportion of transboundary basin areas within Bangladesh, Honduras, and Uganda[15] on which a formal agreement exists, using this simplified definition of cooperation.

A wide range of proportions exists between the three countries. Uganda has the greatest extent of area with an agreement. The *Agreement on the Nile River Basin Cooperative Framework* (CFA) (2010) covers the Nile basin in Uganda and includes hydrologically connected groundwaters in its definition of what is included in the Nile River system; this single agreement accounts for the high percentage of coverage since the Nile River basin and connected

Table 2.1 Percent transboundary area with an existing agreement for Bangladesh, Honduras, and Uganda

Country	*(1) Percent transboundary basin area with an existing agreement*	*(2) Percent transboundary aquifer area with an existing agreement*[a]	*(3) Percent total transboundary area with an existing agreement*
Bangladesh	90	0	49
Honduras	32	0	26
Uganda	98	83	97

[a] This includes aquifers that are part of agreements on overlying river basins, that is, those attempting to manage surface water and hydrologically connected groundwater conjunctively. This is recommended by the UNECE Model Provisions on Transboundary Groundwaters (UNECE 2014).

transboundary aquifers are all part of the Nile River system and subject to the CFA. Comparatively, in Bangladesh, the *1996 Ganges Waters Treaty* also covers the majority of the transboundary basin area within Bangladesh. Because of this agreement,[16] Bangladesh's result for the percent of transboundary basin area with an agreement (column 1) is very similar in magnitude to the result in Uganda, as much like the Nile basin, the Ganges basin constitutes the majority of the nation state's area.

Existing agreements for shared waters may help indicate the predilection for future cooperation; however, as shown by comparing the results, the quality and degree of transboundary cooperation are not exposed by this simplistic interpretation. Furthermore, this lacks the ability to capture whether the agreement is functioning and effective, which could lead to the inclusion of 'paper tigers' – or agreements that have been signed but not implemented. This also fails to capture cooperation that exists without a formal agreement. For example, the Goascorán basin in Honduras has a functioning river basin organization, but no treaty exists between Honduras and its co-riparians. Therefore, this initial example illustrates the need for identifying an explicit definition for cooperation and a means to capture this understanding through a measurement, whether an indicator, an index, or a framework.

Ultimately, the example demonstrates that measuring cooperation through the presence of an agreement, while a common practice in the literature, does not fully represent the state of cooperation. Before we can measure cooperation, we must clearly define what we want to measure. This presents the first of the two challenges this book addresses: How can transboundary water cooperation be defined clearly and consistently?

The first challenge is that there is no single accepted definition for transboundary water cooperation (Tarlock 2015; Saruchera and Lautze 2015). Cooperation is generally defined as coordination and collaboration between states that achieve a common interest (Leb 2015; Zartman 2008). In the operationalization of this definition, complications arise, as the term 'transboundary water cooperation' can take on a wide range of forms. This presents difficulties when trying to measure cooperation in a consistent and comparable manner. Much of the literature focuses on measuring inputs to cooperation rather than the outcomes of cooperation (Tarlock 2015). For example, formal cooperation is often defined by the existence of a treaty or agreement, as demonstrated. This assessment does not reflect the results of cooperation, whether it was effective in improving water security or meeting shared interests, nor does it determine if the treaty or agreement is still operational and not a 'paper tiger.' The following Table 2.2 presents several examples of definitions

Table 2.2 Selection of definitions and measurement criteria for transboundary water cooperation

Term	*Framework, project, study*	*Definition/criteria*
Arrangement for water cooperation	Step-by-Step Monitoring Methodology for Indicator 6.5.2 (UNECE Water Secretariat 2020)	"A bilateral or multilateral treaty, convention, agreement, or other formal arrangement, such as a memorandum of understanding, between riparian countries that provides a framework for cooperation on transboundary water management. Agreements or other kinds of formal arrangements may be interstate, intergovernmental, inter-ministerial, interagency or between regional authorities."
Operational cooperation		The following criteria must be met for an agreement or other formal arrangement to have operational cooperation: • Joint body, joint mechanism, or commission • Regular formal communication • Joint or coordinated water management plan or joint objectives • Regular exchange of data and information
Arrangement for operational cooperation	Transboundary Water Cooperation and the Sustainable Development Goals (Sindico 2016)	An arrangement is considered operational if there are "regular meetings of the riparian countries to discuss IWRM and/or exchange information."
Transboundary water cooperation	Policy Guidance Note on the Benefits of Transboundary Water Cooperation (UNECE 2015)	"Effective cooperation between two or more countries sharing a transboundary river, lake, or aquifer." It includes a continuum of modes:

(*continued*)

Table 2.2 Cont.

Term	*Framework, project, study*	*Definition/criteria*
		• Information sharing • Coordination • Collaboration • Joint action It can be defined as "any action or set of actions by riparian states that leads to the enhanced management or development of the transboundary water body to their mutual satisfaction."
Transboundary water cooperation	Swiss Position Paper on Water in the Post-2015 Agenda (Schweizerische Eidgenossenschaft 2013) A Post-2015 Global Goal for Water: Synthesis of key findings and recommendations from UN-Water (UN-Water 2014)	Measured by the presence of a transboundary water agreement.
Transboundary cooperation	Draft 3rd State of the Nile Report Indicators (AbuZeid 2016)	Calculated on an individual country basis by dividing the number of riparian countries that the country in question has cooperative agreements with by the total number of riparians to the country in question.
Institutional vulnerability	Mapping the Resilience of International River Basins to Future Climate Change Induced Water Variability (De Stefano et al. 2010)	A lack of a treaty or river basin organization.
Institutional resilience		Higher capacity to deal with climate-change-induced water variability in an international basin. Mechanisms used to rate a treaty's capacity for resilience: • Presence of a water treaty • Allocation mechanism • Mechanism for variability management • Conflict resolution mechanism • River basin organization
Facilitating cooperation and averting conflict	The effectiveness of treaty design in addressing water disputes (Mitchell and Zawahri 2015)	Treaties are more effective when they have mechanisms for: • River basin organizations • Monitoring • Conflict resolution • Enforcement • Information exchange

Table 2.2 Cont.

Term	*Framework, project, study*	*Definition/criteria*
Water Cooperation	Measuring Transboundary Water Cooperation: Learning from the Past to Inform the Sustainable Development Goals (Saruchera and Lautze 2015)	Six indicators are recommended to measure the level of cooperation in shared waters: • "Conclusion of a basin-wide transboundary water agreement • Reference to transboundary cooperation in national water legislation • Inclusive completion of a basin plan • Evidence of regular data exchange • Harmonized standards and units of measurement • Riparian financing for transboundary institutional activities"
Governance capacity (transboundary cooperation)	Transboundary Water Assessment Program (TWAP) (UNEP-DHI 2016)	Two indicators within the TWAP project capture transboundary cooperation:
	Assessment of Transboundary River Basins for Potential Hydro-political Tensions (De Stefano et al. 2017)	Legal Framework – considers key principles in water treaties: • Equitable and reasonable use • Not to cause significant harm • Environmental protection • Cooperation and information exchange • Notification, consultation, or negotiation • Peaceful dispute settlement • Ratification of 1997 UN Watercourses Convention or 1992 UNECE Water Convention Hydropolitical Tension – considers the institutional resilience: • Existence of treaty • Allocation mechanism • Flow variability mechanism • Conflict resolution mechanism • River basin organization

and methods for measuring transboundary water cooperation. Cooperation can range from procedural to substantive criteria, paper to practice-based, objective to subjective, and normative to place-based, among other potential differences. These differences in how transboundary water cooperation is conceptualized can challenge consistent monitoring efforts in accurately reflecting the extent of cooperation.

Challenge 2: How can transboundary water cooperation be effective?

The variability in defining transboundary water cooperation has led to efforts to measure cooperation based on outputs of cooperation – ultimately, that cooperation should be effective and produce measurable benefits. The term *effective*, however, can also have a range of meanings depending on the focus or lens through which *effective* is defined in the context of a particular study. This presents the second challenge addressed in this book – defining effectiveness. In some instances, *effective* may not be defined explicitly, creating space for multiple interpretations. For example, is cooperation effective if it prevents conflict, or is it effective if it increases the available water supply for a riparian? *Effectiveness* is a subjective measure for evaluating transboundary cooperation, and like cooperation itself, it needs to have a clear and consistent definition. Table 2.3 presents a selection of different definitions and conceptions of *effectiveness*.

Table 2.3 Selection of definitions for 'effective' in a cooperation or governance context

Term	*Framework, project, study*	*Definition/criteria*
Effective cooperation	Promoting effective water management cooperation among riparians (Tarlock 2015)	Provides measurable benefits to riparians that promote water security. It is a multistage continuum of relations between riparians.
Effective water governance	Effective water governance (Rogers and Hall 2003)	Needs to have an enabling environment, which facilitates efficient private and public sector initiatives, with stakeholder involvement. To make effective water governance: • Raise political will • Practice IWRM • Develop and reform water institutions • Financial and economic support
Effectiveness	The effectiveness of treaty design in addressing water disputes (Mitchell and Zawahri 2015)	The ability of treaties to manage water disputes. An institution is effective if it can address disputes between riparians peacefully and encourage interstate cooperation.
Effectiveness[a]	The Concept of Regime "Effectiveness" (Underdal 1992)	Measured by comparing against a standard of success or accomplishment. This could be compared in terms of *relative improvement* or *collective optimum*. By considering the two together, a two-dimensional conceptualization of effectiveness can be developed.

Table 2.3 Cont.

Term	*Framework, project, study*	*Definition/criteria*
Effectiveness[b]	Effectiveness of international environmental regimes: Existing knowledge, cutting edge themes, and research strategies (Young 2011)	The extent to which a regime "contributes to solving or mitigating the problems that motivate" the regime's creation. Two conceptions of effectiveness: • Outputs or regulations to move a regime from paper to practice or changes in behavior relative to the problem • Assessment of performance relative to the idea of an ideal outcome Measurable as the difference between the actual performance and the regime-less response divided by the difference of the ideal outcome and the regime-less response.
Treaty effectiveness or treaty resilience	Climate change, conflict, and cooperation: Global analysis of the effectiveness of international river treaties in addressing water variability (Dinar et al. 2015)	Ability to promote cooperation and reduce conflict. Variables influencing treaty effectiveness: • Number of mechanisms • Self-enforcement mechanism • Enforcement mechanism • Adaptability mechanism
Institutional effectiveness	Conflict and cooperation along international rivers: Crafting a model of institutional effectiveness (Berardo and Gerlak 2012)	Conditions under which institutions foster meaningful cooperation. The capacity to perform tasks for which the institution was designed, at two levels: • Interstate agreement • Level of process design elements
Effective water sharing	*Guidelines for Development of Effective Water Sharing Agreements* (Environmental and Water Resources Institute 2012)	"An effective agreement can facilitate adequate planning, conservation, utilization, development, management, and control of water resources on a water basin basis, in a manner that is reasonable and equitable under the circumstances and that causes no significant harm to most other parties." Four guiding principles to effective transboundary water sharing: • Coordination and cooperation • Interdisciplinary analysis • Watershed and river basin planning • Adaptive management

(*continued*)

Table 2.3 Cont.

Term	*Framework, project, study*	*Definition/criteria*
Treaty effectiveness	International River Treaty Effectiveness (Brochmann, Hensel, and Tir 2012)	Preventing future river disputes, which may be conditioned by the treaty emergence process. Ability to manage conflict.

[a] Does not refer to effectiveness of water cooperation, but rather defines effectiveness for cooperation in general.
[b] Not specific to water cooperation refers more broadly to environmental regimes.

These are the two challenges at the core of the problem in monitoring, developing, and evaluating transboundary water cooperation that this book seeks to address. How can we create an objective method for measuring transboundary water cooperation based on a clear and specific definition of what is effective? To address this problem, this book addresses these two challenges in two different sections – Parts II and III. Part II answers the question: How is transboundary water cooperation defined? While Part III tackles the question: What makes transboundary water cooperation effective?

Cooperative efforts over shared waters can take many forms, place matters in developing and continuing cooperation between riparians. Climate, hydrology, sociopolitical environments, and particularly political will play a role in the establishment and persistence of cooperation. Water governance and transboundary cooperation should reflect the place and unique context of a basin or aquifer (Giordano and Shah 2014; Jensen 2013). A prescriptive or normative method for defining what transboundary water cooperation looks like may restrict cooperation from developing in a way that best suits the basin, aquifer, or political situation. Further, it has the potential to restrict flexibility in cooperative arrangements, which is needed to respond to changing conditions (McCaffrey 2003). Therefore, this book argues that there is a need for a clear definition of effective transboundary water cooperation based on practical experience, which can be applied to develop a method for evaluating the extent of transboundary cooperation.

Conclusion and book overview

This chapter introduces the global recognition of transboundary water cooperation as a tool for improved governance and management of transboundary surface and groundwaters. While there is a long history of cooperation, it has risen to the global stage in the last half century, most recently with the inclusion of transboundary water cooperation within the Sustainable Development Goals. Nevertheless, with this increased prominence, there are two main challenges that can impact the development of cooperative arrangements. The first is that there is no single agreed definition for transboundary water cooperation. Most measures of transboundary water cooperation focus on agreements, particularly the number of agreements or their content. However, this, generally, does not measure or capture how actors behave in a cooperative process nor whether the behavior had the intended consequences or outcomes (Daoudy and Kistin 2008). Therefore, we need further research on measures of the quality and outcomes of cooperative processes. This leads to the second challenge: there is no clear, consistent definition of effective transboundary water cooperation in the literature. This book explores these two challenges and develops the methodology based on a further research suggestion by De Stefano et al. (2017, 45) that

> It could be interesting to conduct a global survey among water practitioners in transboundary river basins to evaluate strengths and shortcomings of the current institutional cooperation frameworks. Despite the limitations and subjectivity of such a survey, it would add a new dimension to the analysis, complementing existing datasets that are based on documentary or modeling analyses.

The outcomes of this book utilize data from existing literature on transboundary water cooperation and effectiveness, and it incorporates data from over 50 interviews with academics and practitioners working in transboundary water cooperation from all regions of the world.

To address these challenges, this book is divided into three parts. Part I includes this chapter and Chapter 3; it sets the stage for the two challenges. Chapter 3, in particular, uses the Sustainable Development Goal Indicator on transboundary water cooperation to demonstrate the complication and confusion that arises from a lack of a clear and consistent definition of transboundary water cooperation. Furthermore, it demonstrates how many of the frameworks for measuring and tracking progress toward cooperation are, in fact, measuring whether cooperation is effective or not. Part II follows and addresses the first challenge: How is transboundary water cooperation defined? Within this part, Chapters 4, 5, and 6 explore the literature on cooperation, the key elements that compose definitions of cooperation, and finally present a definition and framework that incorporates this survey for the literature and a synthesis of the human knowledge on transboundary water cooperation. However, having a definition of cooperation is not enough. Part III addresses this final challenge of determining what makes transboundary water cooperation effective. Chapters 7, 8, and 9, in Part III, dive into the literature on effectiveness, determine the aspects that make cooperation effective, explore examples of basins with and without effective cooperation, and finally develop a methodology for evaluating the effectiveness of a transboundary water cooperative process.

Notes

1 The following sections of Chapter 2 are reproduced (with revisions and updates) from *Measuring transboundary water cooperation: options for Sustainable Development Goal Target 6.5* (Global Water Partnership, 2017) written by Melissa McCracken for the Global Water Partnership (GWP) for its Technical Committee Background Papers series: Transboundary surface and groundwaters; Cooperation in the context of international water law; and Challenge 1: Multiple understandings of transboundary water cooperation.

2 Please keep in mind that this book focuses on transboundary water cooperation and effective transboundary water cooperation. Therefore, I use cooperation and effective cooperation, without the "transboundary water" modifier, to still mean transboundary water cooperation or effective transboundary water cooperation, respectively. I will note if I am referring to a broader sense of either cooperation or effective/effectiveness that are unlinked from transboundary water.

3 This thesis was prominent in the early 1990s and has seen a resurgence with the threat of climate change. It argues that water resource scarcity is a cause of armed conflict; climate change is predicted to increase variability in precipitation creating water shortage or flooding, which could lead to conflict. However, this theory has been disproven through empirical studies finding that there is more cooperation between states over water than conflict (Wolf, Yoffe, and Giordano 2003).

4 For more concise writing, the author may at points use "basin" in a general sense to refer to both river basins and groundwater basins; a modifier will be used when referring to either surface or groundwater.

5 Al-Sag/Al-Disi Layer, Genevese Aquifer, Guarani Aquifer, Iullemeden Aquifer System, Northwestern Sahara Aquifer System, Nubian Sandstone Aquifer, and Taoudeni/Tanezrouft Aquifer System.

6 Articles 2, 55, and 56 (Leb 2015; United Nations 1945).

7 "Encourages all Member States, the United Nations system and all other actors to take advantage of the Year to promote actions at all levels, including through international cooperation, as appropriate" (United Nations 2010).

8 Article 8 (United Nations 1997).

9 Amendments to Articles 25 and 26 entered into force on 6 February 2003 (United Nations 2016b).

10 UN General Assembly Resolution A/RES/63/124.

11 UN General Assembly Draft Resolution A/C.6/71/L.22.

12 The first treaty known for cooperation over shared water was in 3100 BC between Lagash and Umma, which ended a dispute over irrigation water (Dinar et al. 2007).

13 These datasets were used for calculating areas in the World Cylindrical Equal Area Projection.

14 The decision to use country level calculations, rather than basin level, is so that the values are comparable to the SDG Indicator 6.5.2 calculations in Chapter 3. SDG Indicator 6.5.2 is a country level indicator to align with the other SDGs.

15 These countries were selected as part of a further analysis of SDG Indicator 6.5.2, which is presented in Chapter 3.

16 For both Uganda and Bangladesh, there are additional agreements that exist in the transboundary area.

References

AbuZeid, K. 2016. "Draft 3rd Arab State of the Water Report Indicators." Center for Environment & Development for the Arab Region & Europe (CEDARE), and Arab Water Council.

Agreement on the Nile River Basin Cooperative Framework. 2010. https://nilebasin.org/images/docs/CFA%20-%20English%20%20FrenchVersion.pdf

Behrmann, Christian, and Raya Stephan. 2010. "The UN Watercourse Convention and the Draft Articles on Transboundary Aquifers: The Way Ahead." In *International Conference Transboundary Aquifers: Challenges and New Directions*. Paris. www.tandfonline.com/doi/abs/10.1080/02508060.2011.597345.

Berardo, Ramiro, and Andrea K. Gerlak. 2012. "Conflict and Cooperation along International Rivers: Crafting a Model of Institutional Effectiveness." *Global Environmental Politics* 12 (1): 101–20.

Brochmann, Marit. 2012. "Signing River Treaties—Does It Improve River Cooperation?" *International Interactions* 38 (2): 141–63. https://doi.org/10.1080/03050629.2012.657575.

Brochmann, Marit, Paul R. Hensel, and Jaroslav Tir. 2012. "International River Treaty Effectiveness." 23–26. San Diego. http://citeseerx.ist.psu.edu/viewdoc/download?doi=10.1.1.727.2938&rep=rep1&type=pdf.

Committee on Foreign Relations. 2011. "Avoiding Water Wars: Water Scarcity and Central Asia's Growing Importance for Stability in Afghanistan and Pakistan." Washington D.C.: United States Senate. www.foreign.senate.gov/imo/media/doc/Senate%20Print%20112-10%20Avoiding%20Water%20Wars%20Water%20Scarcity%20and%20Central%20Asia%20Afgahnistan%20and%20Pakistan.pdf.

Conca, Ken, Fengshi Wu, and Cigi Mei. 2006. "Global Regime Formation or Complex Institution Building? The Principled Content of International River Agreements." *International Studies Quarterly* 50 (2): 263–85. https://doi.org/10.1111/j.1468-2478.2006.00402.x.

Cooley, John K. 1984. "The War over Water." *Foreign Policy*, 54: 3–26. https://doi.org/10.2307/1148352.

Cresswell, Tim. 2013. *Geographic Thought*. West Sussex, UK: Wiley-Blackwell.

Daoudy, Marwa, and Elizabeth J. Kistin. 2008. "Beyond Water Conflict: Evaluating the Effects of International Water Cooperation." At the 49th Annual Conference of the International Studies Association on 26–29 March 2008 in San Francisco.

De Stefano, Lucia, James Duncan, Shlomi Dinar, Kerstin Stahl, Kenneth Strzepek, and Aaron Wolf. 2010. "Mapping the Resilience of International River Basins to Future Climate Change-Induced Water Variability." Water Sector Board Discussion Paper Series 15. Washington, D.C.: The World Bank.

De Stefano, Lucia, Jacob D. Petersen-Perlman, Eric A. Sproles, Jim Eynard, and Aaron T. Wolf. 2017. "Assessment of Transboundary River Basins for Potential Hydro-Political Tensions." *Global Environmental Change* 45 (July): 35–46. https://doi.org/10.1016/j.gloenvcha.2017.04.008.

Dinar, Ariel, Shlomi Dinar, Stephen McCaffrey, and Daene Mckinney. 2007. *Bridges over Water*. Singapore: World Scientific Publishing Co.

Dinar, Shlomi, David Katz, Lucia De Stefano, and Brian Blankespoor. 2015. "Climate Change, Conflict, and Cooperation: Global Analysis of the Effectiveness of International River Treaties in Addressing Water Variability." *Political Geography* 45 (March): 55–66. https://doi.org/10.1016/j.polgeo.2014.08.003.

Environmental and Water Resources Institute. 2012. *Guidelines for Development of Effective Water Sharing Agreements*. Reston, Virginia: American Society of Civil Engineers.

Giordano, Mark, Alena Drieschova, James A. Duncan, Yoshiko Sayama, Lucia De Stefano, and Aaron T. Wolf. 2013. "A Review of the Evolution and State of Transboundary Freshwater Treaties." *International Environmental Agreements: Politics, Law and Economics* 14 (3): 245–64. https://doi.org/10.1007/s10784-013-9211-8.

Giordano, Mark, and Tushaar Shah. 2014. "From IWRM Back to Integrated Water Resources Management." *International Journal of Water Resources Development* 30 (3): 364–76. https://doi.org/10.1080/07900627.2013.851521.

Gleick, Peter H. 1993. "Water and Conflict: Fresh Water Resources and International Security." *International Security* 18 (1): 79–112.

Hamner, Jesse, and Aaron T. Wolf. 1997. "Water: I. Patterns in International Water Resource Treaties: The Transboundary Freshwater Dispute Database." *Colorado Journal of International Environmental Law and Policy* 1997: 157–219.

IGRAC and UNESCO-IHP. 2015. "Transboundary Aquifers of the World Map 2015." International Groundwater Resources Assessment Centre. 2015. www.un-igrac.org/resource/transboundary-aquifers-world-map-2015.

International Law Association. 2004. "Berlin Conference on Water Resources Law." International Law Association. http://internationalwaterlaw.org/documents/intldocs/ILA_Berlin_Rules-2004.pdf.

Jensen, Kurt Mørck. 2013. "Swimming Against the Current: Questioning Development Policy and Practice." *Water Alternatives* 6 (2): 276.

Leb, Christina. 2013. "The UN Watercourses Convention: The Éminence Grise behind Cooperation on Transboundary Water Resources." *Water International* 38 (2): 146–55. https://doi.org/10.1080/02508060.2013.781474.

———. 2015. "One Step at a Time: International Law and the Duty to Cooperate in the Management of Shared Water Resources." *Water International* 40 (1): 21–32. https://doi.org/10.1080/02508060.2014.978972.

McCaffrey, Stephen C. 2003. "The Need for Flexibility in Freshwater Treaty Regimes." *Natural Resources Forum* 27 (2): 156–62. https://doi.org/10.1111/1477-8947.00050.

McCracken, Melissa, and Aaron T. Wolf. 2019. "Updating the Register of International River Basins of the World." *International Journal of Water Resources Development* 35 (5): 732–82. https://doi.org/10.1080/07900627.2019.1572497.

Mitchell, Sara McLaughlin, and Neda A. Zawahri. 2015. "The Effectiveness of Treaty Design in Addressing Water Disputes." *Journal of Peace Research* 52 (2): 187–200. https://doi.org/10.1177/0022343314559623.

Rogers, Peter, and Alan W Hall. 2003. *Effective Water Governance*. Technical Committee Background Papers 7. Stockholm: Global Water Partnership.

Sadoff, Claudia W, and David Grey. 2002. "Beyond the River: The Benefits of Cooperation on International Rivers." *Water Policy* 4 (5): 389–403. https://doi.org/10.1016/S1366-7017(02)00035-1.

Saruchera, D., and J. Lautze. 2015. "Measuring Transboundary Water Cooperation: Learning from the Past to Inform the Sustainable Development Goals." International Water Management Institute (IWMI). www.iwmi.cgiar.org/publications/iwmi-working-papers/iwmi-working-paper-168/.

Schweizerische Eidgenossenschaft. 2013. "Swiss Position Paper on Water in the Post-2015 Agenda." Schweizerische Eidgenossenschaft. www.swisswaterpartnership.ch/wp-content/uploads/2014/07/2013.12.20_Swiss_Position_on_Water_in_Post2015_Agenda.pdf.

Sindico, Francesco. 2016. "Transboundary Water Cooperation and the Sustainable Development Goals." Paris: UNESCO-IHP.

Swain, Ashok. 2015. "Water Wars." In *International Encyclopedia of the Social & Behavioral Sciences*, edited by James Wright, Second Edition, 443–47. Amsterdam: Elsevier.

Tanzi, Attila. 2000. "The Relationship between the 1992 UNECE Convention on the Protection and Use of Transboundary Watercourses and International Lakes and the 1997 UN Convention on the Law of the Non-Navigational Uses of International Watercourses." Geneva: UNECE Convention on the Protection and Use of Transboundary Watercourses and International Lakes. www.unece.org/fileadmin/DAM/env/water/publications/documents/conventiontotal_Eng_final.pdf.

Tarlock, Dan. 2015. "Promoting Effective Water Management Cooperation Among Riparian Nations." *TEC Background Papers*, no. 21. www.gwp.org/Global/ToolBox/Publications/Background%20papers/GWP_TEC21_web.pdf.

TFDD. 2021. "Transboundary Freshwater Dispute Database." College of Earth, Ocean and Atmospheric Sciences, Oregon State University. 2021. https://transboundarywaters.science.oregonstate.edu/.

UN General Assembly. 2013. *68/118 The Law of Transboundary Aquifers*. Vol. A/RES/68/118. www.researchgate.net/profile/Alfonso_Rivera2/publication/271829536_A_synthesis_of_knowledge_of_the_Milk_River_Transboundary_Aquifer_Alberta_Canada-Montana_USA/links/54d278c10cf25017917e4c8e.pdf.

UN Watercourses Convention. n.d. "UN Watercourses Convention and the UNECE Water Convention: User's Guide Fact Sheet Series Number 12." UN Watercourses Convention. Accessed November 13, 2016. www.unwatercoursesconvention.org/documents/UNWC-Fact-Sheet-12-Relationship-with-UNECE-Water-Convention.pdf.

Underdal, Arild. 1992. "The Concept of Regime 'Effectiveness.'" *Cooperation and Conflict* 27 (3): 227–40.

UNECE. 2004. "What Is It? Why It Matters? The 1992 UNECE Convention on the Protection and Use of Transboundary Watercourses and International Lakes." UNECE. www.unece.org/fileadmin/DAM/env/water/documents/brochure_water_convention.pdf.

———. 2014. "Model Provisions on Transboundary Groundwaters." United Nations Economic Commission for Europe. www.unece.org/fileadmin/DAM/env/water/publications/WAT_model_provisions/ece_mp.wat_40_eng.pdf.

———. 2015. *Policy Guidance Note on the Benefits of Transboundary Water Cooperation*. Geneva: UNECE. www.unece.org/fileadmin/DAM/env/water/publications/WAT_Benefits_of_Transboundary_Cooperation/ECE_MP.WAT_47_PolicyGuidanceNote_BenefitsCooperation_1522750_E_pdf_web.pdf.

———. 2016a. "The Global Opening of the 1992 Water Convention." UNECE. www.unece.org/fileadmin/DAM/env/water/publications/WAT_The_global_opening_of_the_1992_UNECE_Water_Convention/ECE_MP.WAT_43_Rev1_ENGLISH_WEB.pdf.

———. 2016b. "Water Convention." United Nations Economic Commission for Europe. 2016. www.unece.org/env/water.

UNECE Water Secretariat. 2020. "Step-by-Step Monitoring Methodology for SDG Indicator 6.5.2 2020." UN-Water. www.unwater.org/publications/step-step-methodology-monitoring-transboundary-cooperation-6-5-2/.

UNEP-DHI. 2016. "Transboundary River Basins: Status and Trends."Volume 3: River Basins. Transboundary Waters Assessment Programme. Nairobi, Kenya: United Nations Environment Programme.

UNESCO-IHP and UNEP. 2016. "Transboundary Aquifers and Groundwater Systems of Small Island Developing States: Status and Trends: Summary for Policy Makers." Nairobi, Kenya: United Nations Environment Programme.

United Nations. 1945. *Charter of the United Nations*. https://treaties.un.org/doc/publication/ctc/uncharter.pdf.

———. 1997. *Convention on the Law of Non-Navigational Uses of International Watercourses. A/RES/51/229*. http://heinonline.org/hol-cgi-bin/get_pdf.cgi?handle=hein.journals/tilj33&section=22.

———. 2010. *Resolution Adopted by the General Assembly on December 2010: International Year of Water Cooperation, 2013. A/RES/65/154*.

———. 2016a. "Convention on the Law of the Non-Navigational Uses of International Watercourses – Status." United Nations Treaty Collection. 2016. https://treaties.un.org/Pages/ViewDetails.aspx?src=IND&mtdsg_no=XXVII-12&chapter=27&clang=_en.

———. 2016b. "Convention on the Protection and Use of Transboundary Watercourses and International Lakes." United Nations Treaty Collection. 2016. https://treaties.un.org/Pages/ViewDetails.aspx?src=TREATY&mtdsg_no=XXVII-5&chapter=27&clang=_en.

———. 2016c. "The Law of Transboundary Aquifers - Seventy-First Session." Sixth Committee (Legal) – UN General Assembly. 2016. www.un.org/en/ga/sixth/71/transboundary_aquifers.shtml.

United Nations Economic Commission for Europe. 2013. *Convention on the Protection and Use of Transboundary Watercourses and International Lakes Amended*. www.unece.org/fileadmin/DAM/env/documents/2013/wat/ECE_MP.WAT_41.pdf.

UN-Water. 2014. "A Post-2015 Global Goal for Water: Synthesis of Key Findings and Recommendations from UN-Water." UN-Water. www.un.org/waterforlifedecade/pdf/27_01_2014_un-water_paper_on_a_post2015_global_goal_for_water.pdf.

Wolf, Aaron T. 1998. "Conflict and Cooperation along International Waterways." *Water Policy* 1 (2): 251–65. https://doi.org/10.1016/S1366-7017(98)00019-1.

Wolf, Aaron T. 2007. "Shared Waters: Conflict and Cooperation." *Annual Review of Environment and Resources* 32 (1): 241–69. https://doi.org/10.1146/annurev.energy.32.041006.101434.

Wolf, Aaron T., Jeffrey A. Natharius, Jeffrey J. Danielson, Brian S. Ward, and Jan K. Pender. 1999. "International River Basins of the World." *International Journal of Water Resources Development* 15 (4): 387–427. https://doi.org/10.1080/07900629948682.

Wolf, Aaron T., Kerstin Stahl, and Marcia F. Macomber. 2003. "Conflict and Cooperation within International River Basins: The Importance of Institutional Capacity." *Water Resources Update* 125 (2): 31–40.

Wolf, Aaron T., Shira B. Yoffe, and Mark Giordano. 2003. "International Waters: Identifying Basins at Risk." *Water Policy* 5 (1): 29–60. https://doi.org/10.2166/wp.2003.0002.

Wouters, Patricia. 2013. "International Law–Facilitating Transboundary Water Cooperation." *Published by the Global Water Partnership, TEC Background Papers*, no. 17. http://papers.ssrn.com/sol3/papers.cfm?abstract_id=2363809.

Yamada, Chusei. 2011. "Codification of the Law of Transboundary Aquifers (Groundwaters) by the United Nations." *Water International* 36 (5): 557–65. https://doi.org/10.1080/02508060.2011.597345.

Yoffe, Shira B., and Aaron T. Wolf. 1999. "Water, Conflict and Cooperation: Geographical Perspectives." *Cambridge Review of International Affairs* 12 (2): 197–213. https://doi.org/10.1080/09557579908400256.

Young, Oran R. 2011. "Effectiveness of International Environmental Regimes: Existing Knowledge, Cutting-Edge Themes, and Research Strategies." *Proceedings of the National Academy of Sciences* 108 (50): 19853–60.

Zartman, I. William, Ed. 2008. "Concept: Cooperation." *PIN Points* 30: 5–7.

Zeitoun, Mark, and Naho Mirumachi. 2008. "Transboundary Water Interaction I: Reconsidering Conflict and Cooperation." *International Environmental Agreements: Politics, Law and Economics* 8 (4): 297–316. https://doi.org/10.1007/s10784-008-9083-5.

3 Demonstrating the challenges: SDG Indicator 6.5.2

Measuring operational arrangements for transboundary water cooperation[1]

In a general sense, cooperation can be defined as coordination and collaboration between riparians that achieve a common interest resulting in mutual benefits; this occurs with reciprocity and good faith (Leb 2015; Zartman 2008). The duty to cooperate has become an element of customary international law. Out of this duty, international water law has developed to establish procedural and substantive rules to govern the relations between countries over shared waters (Wouters 2013). As discussed in Chapter 2, the most common expression is through international agreements between riparian states, with an overwhelming majority of these agreements in place on shared surface waters. Agreements, such as treaties, are a type of institutional capacity that can add hydropolitical resilience to a shared water system to absorb change and reduce the potential for conflict to occur (Wolf, Yoffe, and Giordano 2003; Wolf 2007). Despite the fact that research shows that cooperation, particularly through institutional mechanisms, can reduce conflict, the field of international waters still lacks an agreed upon, consistent, and clear definition for transboundary water cooperation. This chapter presents how the evaluation of transboundary water cooperation can vary based on different definitions. Multiple conceptualizations can cause varying results when measuring progress or gauge the extent, level, or stage of transboundary water cooperation. Part I of this book, including this chapter, aims to demonstrate why a clear and consistent definition of transboundary water cooperation is needed. In addition, this chapter illustrates the need for a framework to consider the quality, success, or failure – ultimately, the effectiveness – of cooperative processes. These are the two challenges this book, through Parts II and III, addresses.

In order to demonstrate the two challenges, this chapter compares three different methods for defining transboundary water cooperation based on the Sustainable Development Goal (SDG) Indicator 6.5.2. The indicator was chosen to demonstrate these challenges because water is central to sustainable development, and with a significant proportion of the world's population dependent on internationally shared waters, achieving cooperation over transboundary waters, such as measured through SDG Indicator 6.5.2, will aid in accomplishing the other water targets in SDG 6 as well as the targets indirectly related to water.[2] Furthermore, the prominence of the SDG framework

DOI: 10.4324/9781003126249-4

will elevate global awareness of this conceptualization and method for measuring transboundary water cooperation, making it a highly influential definition. This chapter compares the three methods based on this methodology for the indicator, essentially presenting three similar yet different conceptualizations of transboundary water cooperation. These conceptualizations are compared in three different case study countries – Bangladesh, Honduras, and Uganda. The chapter begins with a brief overview of the SDGs and SDG Indicator 6.5.2, then dives into applying the three methodologies in the three countries. It concludes by comparing the three methods and discussing how this comparison demonstrates the need for a clear, consistent definition for transboundary water cooperation.

Sustainable Development Goals

In 2015, the UN General Assembly, including 193 member states, signed resolution A/RES/70/1 *Transforming our world: the 2030 Agenda for Sustainable Development*, which established the SDGs. The 17 SDGs and their 169 targets greatly increase the scale, ambition, and interconnection of the international agenda, particularly when compared to the scope of the Millennium Development Goals (MDG). These goals aim to complete the unfinished MDG agenda to eradicate poverty while also including integrated goals to achieve sustainable – economic, social, and environmental – development. Over the next 15 years (from 2015), the universal SDGs will apply to both developed and developing countries and will only be achieved through an integrated approach that involves multiple sectors, governments, and administrative units at various social and geographic scales. The priorities of the SDGs include poverty elimination and increases in health, education, and food security. These and other social, economic, and environmental development priorities are reflected in the individual goals and targets (see Box 3.1 detailing individual goals).

Box 3.1 The Sustainable Development Goals (United Nations 2015)

Goal 1: End poverty in all its forms everywhere
Goal 2: End hunger, achieve food security and improve nutrition and promote sustainable agriculture
Goal 3: Ensure healthy lives and promote well-being for all at all ages
Goal 4: Ensure inclusive and equitable quality education and promote lifelong learning opportunities for all
Goal 5: Achieve gender equality and empower all women and girls
Goal 6: Ensure availability and sustainable management of water and sanitation for all
Goal 7: Ensure access to affordable, reliable, sustainable and modern energy for all

Goal 8: Promote sustained, inclusive and sustainable economic growth, full and productive employment and decent work for all
Goal 9: Build resilient infrastructure, promote inclusive and sustainable industrialization and foster innovation
Goal 10: Reduce inequality within and among countries
Goal 11: Make cities and human settlements inclusive, safe, resilient and sustainable
Goal 12: Ensure sustainable consumption and production patterns
Goal 13: Take urgent action to combat climate change and its impacts
Goal 14: Conserve and sustainably use the oceans, seas and marine resources for sustainable development
Goal 15: Protect, restore and promote sustainable use of terrestrial ecosystems, sustainably manage forests, combat desertification, and halt and reverse land degradation and halt biodiversity loss
Goal 16: Promote peaceful and inclusive societies for sustainable development, provide access to justice for all and build effective, accountable and inclusive institutions at all levels
Goal 17: Strengthen the means of implementation and revitalize the Global Partnership for Sustainable Development

During the development of the SDGs, there was a global push to ensure the inclusion of a specific goal related to water within the 2030 Agenda. The MDGs had previously included a water-related target[3] under Goal 7: Ensure Environmental Sustainability (United Nations n.d.). However, this was a limited focus on drinking water and sanitation and embedded within the other environmental targets of MDG 7. Several organizations called for water to figure prominently in what would become the SDGs. The 2013 UN-Water Analytical Brief acknowledged the increased awareness of water security and the need to recognize the implication it has for peace and national security in addition to development issues. The report stated, "water security needs to be included in the formulation of the SDGs" (UN-Water 2013). At the High-Level Roundtable Discussion on Water, Peace, and Security, then US Secretary of State Clinton observed the role water plays in ensuring peace and human development (UN-Water 2013). In 2015, the UN Secretary General Advisory Board on Water and Sanitation noted that the importance of water to development deserved the establishment of a new 'global arena' in the UN, and their report specifically noted that better water resources management is needed within and between countries, as well as across sectors (Alexovich 2015). Stakeholders globally also supported the inclusion of a specific water-related goal during the stakeholder consultation process. Findings of the consultation process found that stakeholders had a consistent message to political leaders that it was time to move forward with water (Taylor 2013). The outcome of this global support and recognition of the importance of water to development resulted in incorporating a specific water goal with eight targets (see Box 3.2).

Box 3.2 Sustainable Development Goal 6 (United Nations 2015)

Goal 6: Ensure availability and sustainable management of water and sanitation for all

- 6.1: By 2030, achieve universal and equitable access to safe and affordable drinking water for all
- 6.2: By 2030, achieve access to adequate and equitable sanitation and hygiene for all and end open defecation, paying special attention to the needs of women and girls and those in vulnerable situations
- 6.3: By 2030, improve water quality by reducing pollution, eliminating dumping and minimizing release of hazardous chemicals and materials, halving the proportion of untreated wastewater and substantially increasing recycling and safe reuse globally
- 6.4: By 2030, substantially increase water use efficiency across all sectors and ensure sustainable withdrawals and supply of freshwater to address water scarcity and substantially reduce the number of people suffering from water security
- 6.5: By 2030, implement integrated water resources management at all levels, including through transboundary cooperation as appropriate
- 6.6: By 2020, protect and restore water-related ecosystems, including mountains, forests, wetlands, rivers, aquifers and lakes
- 6.a: By 2030, expand international cooperation and capacity building support to developing countries in water- and sanitation-related activities and programmes, including water harvesting, desalination, water efficiency, wastewater treatment, recycling and reuse technologies
- 6.b: Support and strengthen the participation of local communities in improving water and sanitation management

Global sentiments, along with various position papers for proposed targets for the SDG on water, included greater emphasis on water resources management when compared with the MDGs (Saruchera and Lautze 2015). For example, the outcomes of stakeholder consultations noted that "achieving improved management of water resources is of extremely high importance across disciplines and segments of society and requires long term support" (Taylor 2013). The various position papers put forward for the targets for the designated water goal attempt to address the connections between water and development, such as water and sanitation and hygiene, water resources management or governance, water-related disasters, wastewater management, and water quality (AMCOW (African Ministers' Council on Water) 2013; UNGC (United Nations Global Compact) 2013; United Nations University and United Nations Office for Sustainable Development 2013; UN-Water 2014; Schweizerische Eidgenossenschaft 2013). Of these position papers submitted prior to the agreement on the 2030 Agenda, only two discussed the importance of including transboundary cooperation as a target for water resources management – Swiss Water Partnership and UN-Water (Schweizerische Eidgenossenschaft 2013; UN-Water 2014).

Target 6.5 builds on the outcomes agreed upon in the 2002 Johannesburg Plan that countries should develop integrated water resource management (IWRM) plans, establish institutions, and integrate management plans at the basin scale (UN-Water 2016). The intention is that an IWRM framework will assist in balancing the other water targets and help to enhance the linkages with the other SDGs. To reach Target 6.5, two indicators have been developed to assist in monitoring and assessing its implementation:

- Indicator 6.5.1: Degree of integrated water resources management implementation

This indicator assesses four components of IWRM implementation – policies, institutions, management tools, and financing – through a survey instrument in the manner of the 2012 Status Report on IWRM (United Nations Environment Programme and United Nations Development Programme 2012). The rationale behind this indicator is that it will assist countries in identifying barriers to IWRM implementation, support the attainment of the other water-related targets, and encourage capacity building at the national level (UN-Water 2016).

- Indicator 6.5.2: Proportion of transboundary basin area with an operational arrangement for water cooperation

This indicator directly monitors the part of Target 6.5, "including through transboundary cooperation as appropriate." Monitoring yields a percentage for each country representing the area of surface- and groundwater basin area with an operational arrangement for transboundary cooperation (UN Statistics 2016a; UN-Water 2016). The rationale for the indicator is to encourage countries to develop operational frameworks for transboundary basins, as the indicator illustrates the areas that lack international arrangements over shared waters.

Monitoring and reporting for Goal 6 are primarily by the individual state. However, the national monitoring efforts are supported and coordinated by international organizations under the UN-Water umbrella. For Target 6.5, the recently created interagency Integrated Monitoring Initiative GEMI coordinates and supports countries in collecting data (UN-Water 2016). For Indicator 6.5.1, UNEP through GEMI, under UN-Water, monitors IWRM status, building upon previous surveys on IWRM implementation. UNECE, as the Secretariat for the UNECE Water Convention,[4] and UNESCO-IHP coordinate national monitoring and aggregation of data for Indicator 6.5.2 (UN Statistics 2016a). The inclusion of transboundary cooperation as part of Target 6.5 and monitoring through Indicator 6.5.2 are important as sustainable development will not be achieved without international cooperation over shared waters. SDG Indicator 6.5.2 measuring the extent of transboundary cooperation is the genesis of the two challenges this book addresses. This chapter demonstrates how the application of the indicator exemplifies the two challenges.

Conceptualizing SDG Indicator 6.5.2

SDG Indicator 6.5.2, given its nature as a global scale measurement in an international framework for sustainable development, is a prominent and influential definition of transboundary water cooperation. Therefore, it presents an excellent case study to explore the two challenges this book addresses – defining transboundary water cooperation and determining what makes it effective. To do this, this chapter evaluates three methodologies for defining and measuring transboundary water cooperation in the style of SDG Indicator 6.5.2. It demonstrates how a difference in the definition of transboundary water cooperation can yield significantly different representations in the same country, even using the same data. Therefore, this section takes a closer look at how SDG Indicator 6.5.2 is structured and how three methods for comparison are defined in this chapter.

As introduced above, SDG Indicator 6.5.2 is the "proportion of transboundary basin area with an operational arrangement for water cooperation" (UN-Water 2016). Within this, there are two key terms that we need to define: *arrangement* and *operational*.

Arrangement for water cooperation is defined in the methodology as

> A bilateral or multilateral treaty, convention, agreement or other arrangement, such as [a] memorandum of understanding, between riparian States that provides a framework for cooperation on transboundary water management. Agreements or other kinds of formal arrangements may be interstate, intergovernmental, interministerial, interagency, or between regional authorities.
>
> (UNECE Water Secretariat 2020, 3)

It paints a broad and inclusive definition of arrangement, but it is centered around the existence of an agreement. The inclusivity is perhaps a means to build flexibility into the definition to allow for context-specific methods of transboundary cooperation. However, this may lead to confusion when identifying what is or is not an arrangement. For example, does a country with a ratified treaty over shared waters have an arrangement for water cooperation, or does a country that is party to one of the two international conventions with no agreement between riparians have an arrangement for water cooperation? The answer is that both have an arrangement for water cooperation, which leads to a subsequent question: should these both be given equal consideration when evaluating the indicator? For this chapter, I use the definition of arrangement in SDG Indicator 6.5.2's methodology, defined at the beginning of this paragraph, for consistency between the methods presented in the following sections. Some deviation occurs in Method 3, which is noted in the respective section.

The second term to define, which is arguably the most controversial, is *operational*. This chapter uses the definition of the term *operational* as the variable between the three methods for calculating SDG Indicator 6.5.2. The intent behind this is to illustrate the variability in how *operational* can be defined and how that can significantly alter the outcome of the indicator and subsequently the representation of the occurring transboundary water cooperation. Since *operational* is the variable in this chapter, the term will be defined separately for each method in the subsequent sections. The three methods are:

- **Method 1**: It is the official SDG Indicator 6.5.2 methodology in "Step-by-Step Monitoring Methodology for Indicator 6.5.2" (UNECE Water Secretariat 2020). This method defines an arrangement as operational if it has all four criteria, including a joint management body, joint management plan or objectives, regular data and information exchange, and regular meetings.
- **Method 2**: It adapts Method 1 based on a recommendation in Sindico's *Transboundary Cooperation and the Sustainable Development Goals* (2016), where an arrangement is considered operational if it meets any one of the criteria in Method 1. This creates levels of operational cooperation, depending on how many criteria are fulfilled.
- **Method 3**: It takes an alternative perspective for defining operational and is adapted from Tarlock's *Promoting Effective Water Management Cooperation among Riparians* (Tarlock 2015). This places arrangements within a typology of cooperation, where their operationality is determined by substantive elements and outcomes of cooperation.

These three methods for defining and assessing transboundary water cooperation are tested in three countries. This enables the comparison between the methods to demonstrate how changing the definition of operational yields different results and highlights how a consistent definition for transboundary water cooperation is needed in the field.

Case study locations

Three countries are chosen as case studies to demonstrate the calculations of the SDG indicator based on the three methods of defining an operational arrangement for transboundary water cooperation. The three countries selected are Bangladesh, Honduras, and Uganda. These countries represent several regions and have a varied number and extent of transboundary river basins and aquifers. This section provides a brief introduction to the hydrologic resources and an overview of the transboundary water cooperation in each country, which will provide context for understanding the results of the three methods for calculating the extent of transboundary water cooperation.

Water resources and transboundary water cooperation in Bangladesh

Overview of water resources

With 57 transboundary rivers, Bangladesh is highly dependent on transboundary water resources. Most of these 57 rivers are either tributaries or distributaries of the Ganges, Brahmaputra, or Meghna rivers and are part of the large transboundary Ganges-Brahmaputra-Meghna River Basin, as shown in Figure 3.1.[5] In addition to the rivers that are part of this system, Bangladesh also shares several other transboundary river basins with both India and Myanmar: the Fenney, Karnaphuli, Muhuri, and Naaf river basins (TFDD 2021). Given Bangladesh's geography, it is the most downstream riparian for these surface water systems, and about 90% of the flow originates from outside of Bangladesh (Ahmed and Roy 2007; Bhuiyan and Hossain 2006; FAO 2014). Despite having some of the largest annual discharges, the temporal variability of monsoonal precipitation within the catchment areas is such that the discharge is unevenly distributed throughout the year, with about 85% of the total flow entering Bangladesh between June and October (FAO 2014). This means a significant deficit exists between the available water resources and demand during

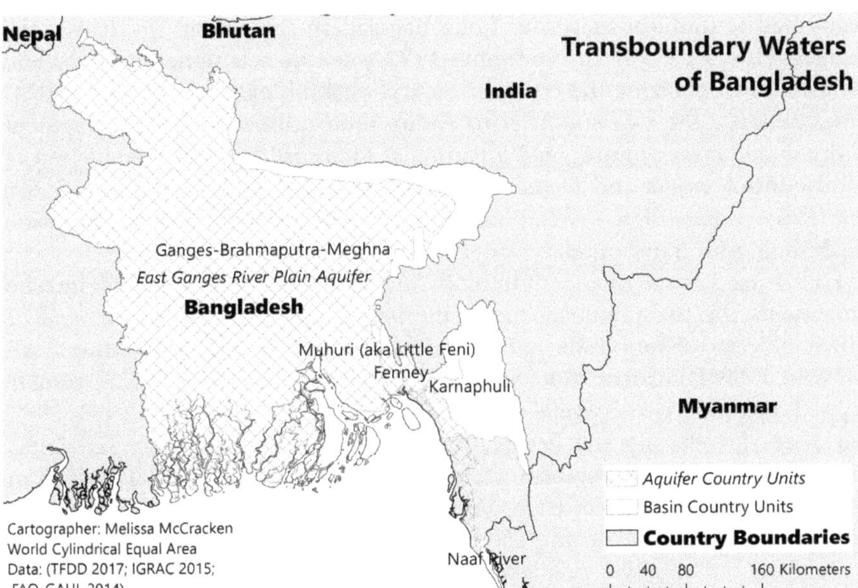

Figure 3.1 Transboundary basin and aquifer country units in Bangladesh.

Source: Reproduced with permission from *Measuring Transboundary Water Cooperation: Options for Sustainable Development Goal Target 6.5* (Global Water Partnership, 2017) written by Melissa McCracken for the Global Water Partnership (GWP) for its Technical Committee Background Papers series.

low flow months and a high potential for flooding and inundation during peak flows. In addition to transboundary surface water, Bangladesh is underlain by a large transboundary aquifer system – the East Ganges River Plain Aquifer (IGRAC and UNESCO-IHP 2015). As much of the country is composed of sedimentary and alluvial deposits from the Ganges-Brahmaputra-Meghna River Basin, the aquifer underlying the country is mostly unconfined and shallow (Ahmed and Roy 2007). Groundwater is available in reasonable quantities and is actively recharged from the surface waters, heavy precipitation, and flood events. In both rural and urban areas, groundwater is the main source of water supply, despite issues with quality, as about 50% of the country is underlain by groundwater bodies that are not suitable to supply drinking water due to arsenic contamination (Ahmed and Roy 2007).

Overview of transboundary cooperation

As a country with such a significant proportion and dependence on transboundary waters, there is a relatively long-standing history of cooperation and disagreement over sharing and managing these waters. The recent origin of conflict and cooperation over the Ganges River begins with the construction of the Farakka Barrage (Rahaman 2006). Through negotiations in the 1960s and 1970s, the Indo-Bangladesh Joint Rivers Commission (JRC) was established to monitor successive India-Bangladesh agreements on sharing the Ganges (Hossain 1998). In November 1972, a statute was signed by India and Bangladesh formalizing the commission and establishing its function; the JRC liaises between the two countries to ensure joint utilization and management of the shared river systems, with a particular focus on the joint formulation of flood control works and technical assessments (*Statute of the Indo-Bangladesh Joint Rivers Commission*, 1972). This commission is the only formal institution for dealing with transboundary water issues, but it is limited in scope.

Following several memorandums of understanding (MOUs) and interim agreements, the two countries signed the first *Ganges Water Agreement of 1977*, which allocated shared water at Farakka and worked toward augmenting flows (Hossain 1998; Rahaman 2006; Sood and Mathukumalli 2011). This agreement expired after five years; several additional MOUs were signed, in 1983, 1984, and 1986, for sharing the dry season flows. In 1996, a 30-year agreement was signed for sharing the Ganges, known as the *Ganges Water Sharing Treaty* (Sood and Mathukumalli 2011). This document is very similar to the 1977 agreement with regard to the allocation of flows at Farakka from 1 January to 31 May but removes discussion of flow augmentation that was included in the 1977 agreement (Nishat and Faisal 2000). For augmentation to occur, the two countries would need to cooperate with other riparians, particularly Nepal, but no mechanism for addressing other riparians to find a sustainable solution to integrated management of the basin is specified in the treaty. The 1996 agreement includes several of the general principles of international law, including those of equity and no significant harm, while also addressing mechanisms for dispute resolution (Hossain 1998; Rahaman 2006).

Of the many transboundary tributaries in the Ganges-Brahmaputra-Meghna River Basin, only one has significant cooperative efforts. The Teesta River is the fourth largest transboundary river in Bangladesh, and it begins in the northeast of India. Negotiations for a water sharing agreement had been ongoing, and a draft interim agreement was developed in 2011, which would divide the dry season flow between India and Bangladesh (Mirchandani 2016; Suryanarayanan 2010). However, political factors intervened, and the agreement was not signed. Despite renewed hope following elections in 2014, the agreement remains unsigned and unimplemented (Mirchandani 2016). In addition to the Teesta River, the 1985 MOU and 1986 *Summary Record of Discussion of the First Meeting of the Joint Committee of Experts* between India and Bangladesh, there has been proposed cooperation to study and share all transboundary waters to mutual benefit, however, whether this study was completed and the sharing of other waters was/is occurring is not apparent.

The National Water Policy of Bangladesh from 1999 emphasizes collaboration with riparians on relevant aspects of managing transboundary water resources, including groundwater (Zahid and Ahmed 2006). Currently, there is no concrete cooperation between Bangladesh and other riparians on shared aquifers.

Water resources and transboundary water cooperation in Honduras

Overview of water resources

Honduras' water resources differ greatly from those of Bangladesh and Uganda. Instead of the majority of the country's area contributing to one transboundary river basin, the topography of Honduras creates several smaller transboundary watersheds, with only about 20% of the country's land area contributing water runoff to shared surface water systems (TFDD 2021). There are six transboundary river basins as seen in Figure 3.2: Choluteca, Coco/Segovia, Goascorán, Lempa, Motagua, and Negro (TFDD 2021). Catchments in Honduras can be divided into two regions: those draining to the Atlantic Ocean and those draining to the Pacific Ocean. More land area drains to the Atlantic, and these watersheds contribute to a greater percentage of the total surface water flow of Honduras (FAO 2015a). Aquifers in Honduras have not been extensively mapped, with most delineations drawn based on geologic formations (GWP Central America 2015). There are five identified transboundary aquifers: Esquipulas-Ocotepeque-Citalá, Estero Real-Río Negro, Chiquimula-Copán Ruinas, Ostua-Metapán, and Delta del Río Motagua (IGRAC and UNESCO-IHP 2015). Aquifers within the Atlantic coastal areas tend to have higher yields, and in the highland areas, groundwater levels are declining, which is where groundwater is an important resource for irrigated agriculture (FAO 2015a).

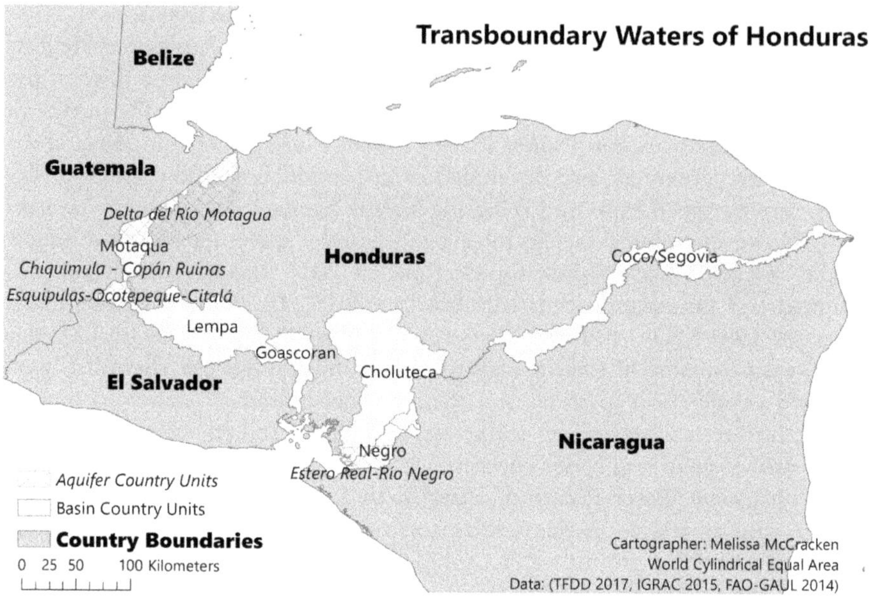

Figure 3.2 Transboundary basin and aquifer country units in Honduras.

Source: Reproduced with permission from *Measuring Transboundary Water Cooperation: Options for Sustainable Development Goal Target 6.5* (Global Water Partnership, 2017) written by Melissa McCracken for the Global Water Partnership (GWP) for its Technical Committee Background Papers series.

Overview of transboundary cooperation

The geography of Honduras' shared waters may be a contributing factor to the limited history of transboundary cooperation between Honduras and its neighboring riparians. Of the Central American countries, only three have national water laws. In Honduras, the 2009 National Water Law establishes a framework of principles and objectives for managing water resources, with the inclusion of IWRM principles (GWP 2016; GWP Central America 2015). However, this law does not address the need to cooperate on transboundary surface- or groundwaters, such as recommending the development of agreements; it does mandate that a representative of the national government should be included in basin councils and transboundary management organizations (GWP Central America 2015; Republica de Honduras 2009).

The only transboundary area with a formal institutional framework is the Upper Lempa River Basin, managed through the Trifinio Plan (United Nations Environment Programme 2007). The Trifinio Plan was developed out of conservation efforts in the 1970s. In 1987, Guatemala, El Salvador, and Honduras developed an updated phase of the Trifinio Plan that included the management and sustainable development of the Motagua, Ulua, and Lempa Rivers; however, all but the Lempa River were dropped from the plan (López 2004). In 1997, the three countries signed a treaty for the execution of the Trifinio Plan, which established the Tri-National Commission, whose role is to administer the plan (López 2004; United Nations Environment Programme 2007). The plan's primary focus is on rural development in general and does not contain provisions for the direct management of the Lempa River (GWP 2016). The scope of the plan has not limited civil society organizations and local authorities to establish plans and activities that address transboundary and integrated water resources management (GWP 2016). For example, the Mancomunidad Trinacional Fronteriza Rio Lempa is a local entity that was developed to fill gaps in the Trifinio Plan and the Tri-National Commission's strategy. It has established its own shared waters policy for transboundary cooperation (Mancomunidad Trinacional Fronteriza Rio Lempa 2016).

The other river basins do not have any formal cooperation. In 2006, the Bi-National Management Group for the Goascorán Basin was established and drafted a Bi-National Management Plan (GWP 2016; MacQuarrie et al. 2013). This group only included local governments and civil society. The Bi-National Management Group was resurrected in 2011 through support from the International Union for Conservation of Nature's (IUCN) Bridge Project (MacQuarrie et al. 2013; Medina 2014). In the Coco/Segovia Basin, UNEP and UN Office for Project Services (UNOPS) partnered with local organizations and municipalities in Honduras and Nicaragua to build watershed management capacity. The project, beginning in 2009, established water management plans for the sub and micro basins in the middle and upper Coco/Segovia basin (UNOPS 2012; Baca, Olivera, Pérez Carrasco, et al. 2012; Baca, Olivera, Leiva, et al. 2012). Neither basin has a specific agreement for cooperative management; this lack of involvement of state actors could be an obstacle to the continued success of the cooperative efforts and create a reliance on international donors (Medina 2014).

Transboundary cooperation is not occurring in shared aquifers. The Trifinio Plan does not mention groundwaters; however, donor projects on aquifers have been completed and recommend that groundwater be included in basin management plans that exist at the local and regional levels (Buch and Guevara 2010).

Water resources and transboundary water cooperation in Uganda

Overview of water resources

Like Bangladesh, Uganda is also dependent on transboundary waters, given that all the country's land area is within a transboundary river basin. The tributaries and lakes within the Nile River Basin comprise most transboundary surface waters; catchment areas that drain into the Lotagipi Swamp and Lake Turkana are the two other transboundary surface waters, as shown in Figure 3.3 (TFDD 2021). Uganda is a landlocked country that is both upstream (to South

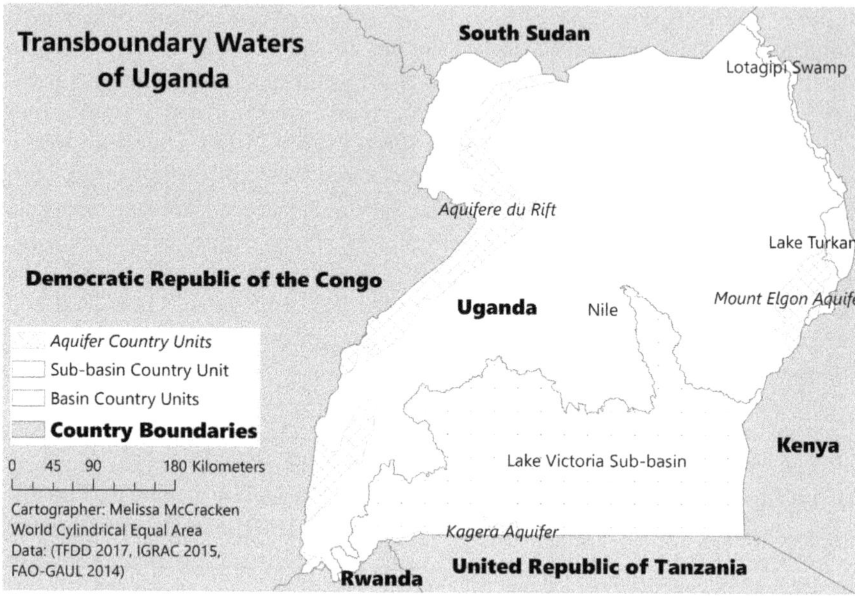

Figure 3.3 Transboundary basin and aquifer country units in Uganda.

Source: Reproduced with permission from *Measuring Transboundary Water Cooperation: Options for Sustainable Development Goal Target 6.5* (Global Water Partnership, 2017) written by Melissa McCracken for the Global Water Partnership (GWP) for its Technical Committee Background Papers series.

Sudan, Sudan, and Egypt) and downstream (to Burundi, Democratic Republic of the Congo (DRC), Kenya, Tanzania, and Rwanda). In terms of groundwater, studies are still ongoing to map groundwater resources, but most of the productive aquifers are found in weathered bedrock or within volcanic formations in mountainous areas (FAO 2015b). Three transboundary aquifers have been mapped: Mount Elgon Aquifer, Kagera Aquifer, and the Aquifere du Rift (IGRAC and UNESCO-IHP 2015). Uganda is generally well-endowed in terms of water resources compared to its needs, with withdrawals in 2008 only reaching a little over 1% of the total annual renewable water resources (FAO 2015b; Nsubuga, Namutebi, and Nsubuga-Ssenfuma 2014). However, the country is heavily dependent on rainfall; therefore, variability in the spatial and temporal distribution of precipitation has significant impacts on water availability and stress (Nsubuga, Namutebi, and Nsubuga-Ssenfuma 2014; Kilimani 2013). Groundwater is a primary source of water supply in rural and arid areas, and generally, extraction rates are less than recharge rates (Kilimani 2013). However, water levels – both surface- and groundwater – have been declining in specific sub-catchments of the Ruizi River, Lake Wamala, and Lake Victoria (DWRM 2011).

Overview of transboundary cooperation

There have been numerous arrangements signed to share the Nile waters, both pre- and post-independence of Uganda. In 1929, Egypt and the British Government, on behalf of Sudan and East African riparians, including Uganda, signed the *Nile Waters Agreement* to protect the interests of Egypt based on 'natural and historic' rights (E. Kasimbazi 2015; Paisley and Henshaw 2013). This agreement gave Egypt overwhelming rights compared to Sudan and the other East African countries. Without a specific timeframe, the countries are bound by the agreement, and post-independence could not renegotiate their positions (E. Kasimbazi 2015). Uganda, as well as Kenya and Tanzania, in accordance with the Nyerere Doctrine on State succession, do not consider themselves bound to the *1929 Agreement* because it was made prior to independence (E. Kasimbazi 2015). The British Government also signed the *1950 Agreement for the Cooperation between the United Kingdom and Egypt*. This agreement established cooperation in meteorological and hydrological surveys (E. Kasimbazi 2015; E.B. Kasimbazi 2010). These two agreements, as well as several others signed by Great Britain on behalf of Uganda, do not acknowledge Uganda's right to use the shared waters (E. Kasimbazi 2015). The 1959 agreement for full utilization of the Nile waters was signed between the independent states of Egypt and Sudan. This agreement differs from the 1929 agreement in that it presumes 'full' use for the two countries. It excludes the other riparians reinforcing the dominant positions of Egypt and Sudan over the other riparians should they claim a share of the Nile waters (E.B. Kasimbazi 2010). Therefore, Uganda and the other upper riparians have found contention with the pre-independence agreements, highlighting the need for a new cooperative arrangement among all the Nile Basin states that recognizes the upper riparians' rights, as well as their evolving socioeconomic situations.

Post-independence, Uganda signed the *Agreement for the Hydrometeorological Survey of Lakes Victoria, Kyoga, and Albert* in 1967, which evaluated the water balance of the lakes with respect to their impact on the flow regime of the Nile (E. Kasimbazi 2015). This project failed due to political disinterest and expired in 1992, but it was replaced by the *Technical Cooperation Committee for the Promotion and Development of the Nile (TECCONILE) Agreement*. This was intended as a transitional arrangement that aimed to contribute to the development of the Nile Basin in an integrated manner (Paisley and Henshaw 2013). It was replaced in March 1999 with the Nile Basin Initiative (NBI), which was also established as a transitional arrangement for sustainable management with the objective of establishing a *Cooperative Framework Arrangement* (CFA) that would be inclusive of all riparians (Salman 2013; Wolf and Newton 2007). The NBI is important, as it represents the first time all Nile Basin countries cooperated to develop and manage the Nile. Principle 15 of the NBI declares all existing agreements inconsistent with the NBI framework null and void, a doctrine that Egypt and Sudan do not agree with (E. Kasimbazi 2015). While the NBI remains the institutional body on the Nile, the CFA was signed in

2010 by Uganda, Tanzania, Rwanda, Kenya, Burundi, and Ethiopia; Egypt, the DRC, South Sudan, and Sudan are yet to sign the agreement, meaning it is not in force. Egypt and Sudan want a provision included in the agreement which states that the water security and current uses and rights of any basin state would not be adversely affected (Salman 2013). As expected, the upper riparian states reject this proposal because it is inconsistent with the goals and vision of the NBI. It is unclear how this will be resolved, given that Egypt and Sudan are parties to the NBI but not signatories to the CFA.

Transboundary cooperation in the Nile Basin also occurs at the sub-basin level. In 1999, the *Treaty for the Establishment of the East African Community* (EAC) was signed between Uganda, Kenya, and Tanzania (Rwanda and Burundi acceded later), which aims to promote sustainable growth and equitable development while protecting the environment (EAC 2000). Per Kasimbazi (2015), the EAC is possibly the most comprehensive regionally binding basis for developing joint strategies for the integrated management of the water resources of Lake Victoria. Under the EAC, Uganda, Kenya, and Tanzania signed the *Protocol for Sustainable Development of Lake Victoria Basin of 2003* (E. Kasimbazi 2015). This protocol promotes cooperation among the lake's riparians, while preventing significant harm to the other Nile Basin states, and it forms the basis for the Lake Victoria Basin Commission, which is responsible for sustainable use and management of the sub-basin.

The two other international basins not connected to the Nile River System – the Lake Turkana Basin and the Lotagipi Swamp – do not have cooperative arrangements occurring with respect to Uganda. In the Lake Turkana Basin, UNEP, Kenya, and Ethiopia have been calling for joint management of the basin and the formulation of a bilateral agreement (Nanni 2016); however, Uganda contributes very minimally to the basin and is unlikely to be involved.

Transboundary groundwaters are only tangentially addressed in aquifers underlying the Nile River; this is through the CFA and some projects funded through the NBI and donor agencies. It is unclear if any formal cooperation or joint management is occurring on these aquifers and those not hydrologically connected to the Nile River.

Method 1: Methodology for SDG Indicator 6.5.2

Currently, UN Statistics has ranked SDG 6.5.2 as a Tier I indicator,[6] meaning that the established methodology and standards are conceptually clear and data is produced by at least 50% of countries where relevant (IAEG-SDG 2016; 2019). Under UN-Water's GEMI, this methodology was developed by the Working Group on Integrated Water Resources Management, which has members from several UN agencies and other organizations and is being coordinated by UNECE Water Secretariat and UNESCO-IHP (UN Statistics 2016b). To identify the "proportion of transboundary basin area with an operational arrangement for water cooperation," UN-Water published a *Step-by-Step Monitoring Methodology for Indicator 6.5.2*, with the most recent revision released in February 2020. This is the methodology that is used for Method 1.[7]

As discussed earlier, how *operational* is defined varies between Methods 1, 2, and 3. With SDG Indicator 6.5.2, if a cooperative arrangement is found to exist on a basin or aquifer, it must meet the following criteria in order to be deemed operational (UNECE Water Secretariat 2020, 3):

- There is a joint body, joint mechanism, or commission (e.g., a river basin commission) for transboundary cooperation.
- There are regular (at least once per year) formal communications between riparian countries in [the] form of meetings (either at the political and/or technical level).
- There is a joint or coordinated water management plan(s), or similar instrument, such as an action plan, common strategy, or joint objectives regarding the status or conditions of the transboundary waters (such as water quality objectives) in place.
- There is regular exchange (at least once per year) of data and information.

These criteria for an arrangement to be considered operational are based on the principles of international law that are codified in the UN Watercourse Convention, UNECE Water Convention, and the Draft Law of Transboundary Aquifers (UN Statistics 2016b). All of the above criteria must be met for the area to count toward the indicator calculation; these do not have to be specified in the arrangement or agreement, but they must be occurring (UN-Water 2017). For example, a criterion could be satisfied by a decision or mechanism by a river basin organization.

To determine if an arrangement meets these criteria, data have been collected from countries through the reporting mechanism under the UNECE Water Convention. The reporting questionnaire for the implementation of the UNECE Water Convention also collects data to track the progress toward transboundary cooperation in the SDGs (Working Group on Integrated Water Resources Management 2016; UN Statistics 2016b). For Method 1, SDG Indicator 6.5.2 is calculated using two ways. First, using the self-reported data from the countries during the second reporting exercise in 2020. Second, the author uses data from the reporting questionnaire[8] that was sent to appropriate contacts within the Global Water Partnership Network in Bangladesh, Honduras, and Uganda. Given the scope of the study, the questionnaire was only sent to in-country representatives; therefore, the responses were not calibrated with the responses from co-riparians to the shared waters. Data from this survey,[9] in addition to available literature and global datasets, including the International Freshwater Treaties and River Basin Organization Databases at Oregon State University,[10] were used to determine the operationality of the identified cooperative arrangements.[11]

The steps taken to calculate the indicator, in both aggregated and disaggregated form, were as follows:

1 Identify transboundary river basins and transboundary aquifers and determine the associated BCUs and ACUs.
2 Identify which basins and aquifers have a cooperative arrangement(s).
3 Determine if the arrangement(s) meets all the criteria to be considered operational.
4 Sum the BCUs with operational arrangements and divide by the total transboundary BCU area to obtain the disaggregated proportion of transboundary river basins that have transboundary cooperation.
5 Sum the ACUs with operational arrangements and divide by the total transboundary ACU area to obtain the disaggregated proportion of transboundary aquifers that have transboundary cooperation.
6 For the SDG Indicator 6.5.2, sum the BCU and ACU areas with operational arrangements and divide by the total area of all the BCUs and ACUs in the country.

For the three case study countries, using the results of the survey, additional literature, and the methodology outlined above, the results are listed in Table 3.1. In addition, the self-reported data from each country during the 2020 reporting exercise is also listed.

Table 3.1 Results for SDG Indicator 6.5.2 using Method 1: Official Methodology (UN-Water 2021)

Data source	*Country*	*1: Proportion of transboundary river basin area with an operational arrangement (%)*	*2: Proportion of transboundary aquifer area with an operational arrangement (%)*	*3: Proportion of total transboundary area with an operational arrangement (%)*
Author	Bangladesh	0	0	0
	Honduras	0	0	0
	Uganda[a]	98	0	90
2020 Reporting Exercise	Bangladesh[b]	–	–	–
	Honduras	0	0	0
	Uganda	98.47	0	84.93

[a] Evaluation was based only on secondary research, as a survey response was not received.
[b] Bangladesh did not provide self-reported data through the formal reporting exercises in 2018 or 2020. See for 2018 reporting results: https://unece.org/environmental-policy/water/transboundary_water_cooperation_reporting. See for 2020 reporting results: www.sdg6data.org/indicator/6.5.2.

Source: Reproduced with permission from *Measuring Transboundary Water Cooperation: Options for Sustainable Development Goal Target 6.5* (Global Water Partnership, 2017) written by Melissa McCracken for the Global Water Partnership (GWP) for its Technical Committee Background Papers series.

Discussion of Method 1

Table 3.1 displays the two sets of results of the Method 1 calculation for SDG Indicator 6.5.2 – from the author calculated data and from the 2020 self-reporting exercise. Columns one and two present the indicator disaggregated for the transboundary river basin area and transboundary aquifer area. Column three presents the aggregated indicator for total transboundary area per country. Bangladesh and Honduras have no ongoing operational cooperative arrangements within transboundary river basins or aquifers. The only area that was calculated as having an operational cooperative arrangement was the transboundary river basin area in Uganda, which contributes to the 90% proportion of the total transboundary area in Uganda that has an operational cooperative arrangement. This 90% value differs from the 84.93% value from the self-reported data due to the differences in the area of the transboundary aquifers.

The 0% value for Bangladesh and Honduras indicates that there is no *operational arrangement* for cooperation occurring in either country, and it also implies that there is no *cooperation* occurring, which is not the case. The structure of the indicator in this method is categorical; to be operational, a cooperative arrangement must meet all the criteria – there is either cooperation or there is not. The binary nature of this conceptualization of operational cooperative arrangements overlooks transboundary cooperation that is occurring without meeting all the criteria. For example, in Bangladesh's BCU for the Ganges-Brahmaputra-Meghna River Basin, only one criterion was not met,[12] that is, preventing the area from contributing to the SDG Indicator calculation. Similarly, the area for the Upper Lempa River in Honduras does not contribute to the indicator, because it lacks formal data and information exchange regarding the shared water resources through the Trifinio Plan and the Tri-National Commission. The *Step-by-Step Monitoring Methodology for Indicator 6.5.2* attempts to address this issue by stating, "the operationality of cooperation is more dynamic as it evolves with the expansion of cooperation. The operationality can be expected to evolve over shorter time frames, and in a year or two, progress could potentially be observed" (UN-Water 2017). However, unless each criterion is satisfied and reported, progress will not be demonstrated in short-term monitoring of the indicator at a global scale. The binary – yes or no – nature of the definition of *operational* will mask any stepwise progress when the data are reported.

During the calculations for Method 1, I assumed the participation of riparians in cooperative arrangements in basins or aquifers that were multilateral. The *Step-by-Step Monitoring Methodology for Indicator 6.5.2* states,

> In situations where more than two riparian countries share a basin, but only some of them have operational cooperation arrangements, the indicator value may mask the gap that a riparian country does not have cooperation arrangements with both its upstream and downstream neighbors.
>
> (UN-Water 2017)

What is unclear in the methodology is whether *all* riparians must be participating in the cooperative arrangement or a specific criterion for the requirements for *operational* to be satisfied. Therefore, I focused the evaluation of the cooperative arrangement on the specific BCU or ACU; if a criterion existed in the BCU or ACU of interest, then it counted toward meeting the requirements for an operational cooperative arrangement, whether or not all riparians to the multilateral basin or aquifer were included. For example, the Indo-Bangladesh JRC and the Joint Committee satisfied the requirement for a joint body, joint mechanism, or commission for transboundary cooperation within the Ganges-Brahmaputra-Meghna River Basin, because they exist within the BCU in Bangladesh, even though only two riparians – India and Bangladesh – of the five are included in the institutions. While the exclusion of riparians may not be ideal, it is important to recognize cooperative efforts that are occurring in multilateral basins or aquifers as multilateral arrangements may be unattainable given political or other context-specific situations. Looking only at the aggregated country-level data, however, misses the entire picture of cooperation in a basin or aquifer. Comparing disaggregated data for BCUs/ACUs within a single basin or aquifer, or aggregating data within a basin, captures an alternative picture of transboundary cooperation in multilateral basins that is overlooked with country-level data (UN-Water 2017).

The final point of discussion for Method 1 emphasizes the definition of *operational* on procedural criteria: a cooperative arrangement, joint body, formal communication, joint water management plan, and data and information exchange. These operational requirements are normative, or they dictate how best to approach transboundary water cooperation. This prescriptive focus on process has the potential to limit adaptation and creativity in cooperative efforts that reflect unique context and place-based specifics. Water governance and transboundary cooperative efforts should allow for context and the sociopolitical environment in which each basin or aquifer is located (Giordano and Shah 2014; Jensen 2013). Flexibility is needed in cooperative agreements, so they are able to respond to changing conditions in the basin or aquifer (McCaffrey 2003). For example, in Saruchera and Lautze's review of transboundary water cooperation indicators, they recommend against requiring the establishment of a river basin organization, as that may encourage a particular type of institutional cooperation that may not be applicable in all contexts (2015). The Columbia River Basin, a well-studied and referenced example of positive transboundary cooperation, does not have an established river basin organization and would not be considered an operational cooperative arrangement under Method 1. Furthermore, the focus on specific procedural criteria may overlook or devalue alternative cooperative efforts that do not 'check all the boxes.' For example, in Honduras, the Goascorán River Basin was found not to have an operational transboundary cooperative arrangement using this method. However, the basin has an alternative cooperative effort in the Bi-National Management Group, which is a multilevel effort that includes local and regional entities and members from public and private sectors. The effort does not have a formal arrangement between Honduras and El Salvador, nor does the organization have support from state actors, which creates a reliance on international donors (Fundación Hondureña de Ambiente y Desarrollo VIDA 2008; MacQuarrie et al. 2013; Medina 2014). Despite this, the cooperative effort is progressing within the political reality and should not be discounted within the scope of the SDG Indicator 6.5.2. Overall, this method presents a good starting place for measuring transboundary cooperation within the constraints of what is feasible for the SDGs and global monitoring efforts; however, users of these data should recognize the limitations of a normative procedural framework for transboundary cooperation, as it does not have the flexibility to capture different contexts, places, sociopolitical situations, or alternative methods of transboundary cooperation.

Method 2: Flexibility in levels of operational cooperation

Method 2 varies slightly from Method 1 and is based on a recommendation put forward by Sindico in *Transboundary Water Cooperation and the Sustainable Development Goals*, a UNESCO-IHP Advocacy Paper (2016). The paper recommends an alteration to an older definition of the indicator that was presented in a UN Statistics document; this document states, "regular meetings of the riparian countries to discuss IWRM **and** exchange of information are required for an arrangement to be defined as 'operational'" (UN Statistics 2016a). Sindico recommends that an arrangement be considered operational if there are "regular meetings of the riparian countries to discuss IWRM **and/or** exchange information" (Sindico 2016). While this older UN Statistics definition of operational is not consistent with the current criteria an arrangement is required to fulfill to be operational, the minor difference between these two statements – changing 'and' to 'and/or' – helps to increase the ability of the SDG Indicator to incentivize transboundary cooperation. Requiring both discussion of IWRM and the exchange of information is a narrow interpretation of what transboundary cooperation is, which would exclude some cooperative arrangements from the indicator (Sindico 2016).

Method 2 uses the same definition of cooperative arrangement and the same criteria for determining if an arrangement is operational as Method 1. The difference between Method 1 and Method 2 is that of 'and' and 'or'. In Method 2, all criteria from Method 1 do not have to be met for a cooperative arrangement to be considered operational. For this method, if any of the criteria are met, the arrangement is considered operational and the area is included in the calculation. There are levels of operational cooperation depending on the number of criteria that the arrangement meets. This method attempts to address the critique of the binary nature of Method 1. By creating levels, the SDG Indicator could help to encourage countries to move toward the goal of transboundary cooperation. The indicator could track progress toward this goal, which through monitoring could reward progress by illustrating growth and attempting to prevent loss of cooperative efforts that may already be in place in a country. For example, a 0% indicator value identifying a lack of cooperation in a country that meets all but one of the criteria could be discouraging and potentially have a negative impact on the current efforts. However, with levels of cooperation, that country's progress could be rewarded, with the indicator serving as an incentive to meet the remaining criteria.

An arrangement is considered operational if it meets at least one of the criteria. Having a formal agreement or similar arrangement will also be included as a criterion for determining levels of operational cooperation for this method. The level of cooperation is determined by the number of criteria met; for example, if two criteria are met, then the area is within Level 2 of operational cooperation. Level 0 will be used to identify those areas that meet none of the criteria, and Level 5 represents the fulfillment of all criteria. The following are the criteria for determining the level of operational cooperative arrangements:

- Existence of a formal cooperative arrangement or agreement.
- Existence of a joint body, mechanism, or commission for transboundary cooperation.
- Regular meetings or formal communication.
- Existence of a joint management plan or objectives.
- Exchange of data and information.

As in Method 1, data from the UNECE Water Convention reporting survey, available literature, and information from global datasets were used to determine which criteria are met. The steps taken to calculate the indicator are the same as the steps described in Method 1. However, instead of three results – one for the extent of transboundary basins, one for the extent of transboundary aquifers, and one for the extent of total transboundary area – there will be a value for every level of operational cooperative arrangement. To calculate this, the areas of the BCUs or ACUs with the same level of cooperation are summed and divided by the respective total area.

The results of Method 2 are included in the following bar charts and tables found in Figures 3.4, 3.5, and 3.6. Each chart shows the percentage of the total transboundary area, transboundary river basin area, and transboundary aquifer area within a specific level of cooperation.

Discussion of Method 2

Figures 3.4 to 3.6 show the results of Method 2 in two ways. The stacked bar charts are a visual representation of the proportion of basin, aquifer, and the total transboundary area within the six levels of operational cooperation. The percentages that compose that bar charts are presented in associated tables. Upon initial comparison of the results of Method 1 and Method 2, it is apparent that more cooperation is occurring than is illustrated in the binary categories of Method 1. Both Honduras and Bangladesh have cooperative efforts that meet several of the criteria for operationality, but these

Method 2: Bangladesh

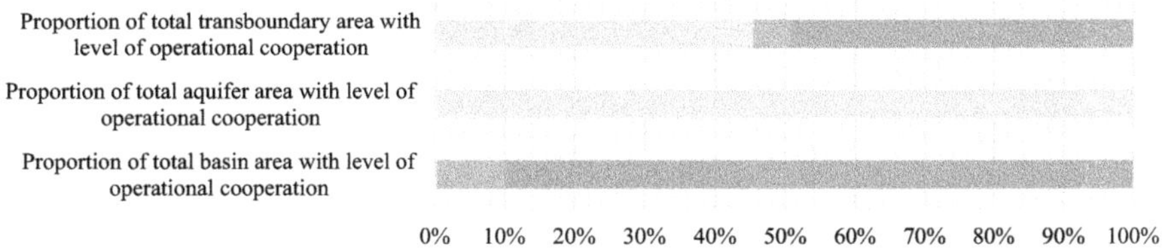

	Proportion of total basin area with level of operational cooperation	Proportion of total aquifer area with level of operational cooperation	Proportion of total transboundary area with level of operational cooperation
Operational Level 0	0%	100%	45%
Operational Level 1	0%	0%	0%
Operational Level 2	0%	0%	0%
Operational Level 3	10%	0%	5%
Operational Level 4	90%	0%	49%
Operational Level 5	0%	0%	0%

Figure 3.4 Results of Method 2 for Bangladesh.

Source: Reproduced with permission from *Measuring Transboundary Water Cooperation: Options for Sustainable Development Goal Target 6.5* (Global Water Partnership, 2017) written by Melissa McCracken for the Global Water Partnership (GWP) for its Technical Committee Background Papers series.

Method 2: Honduras

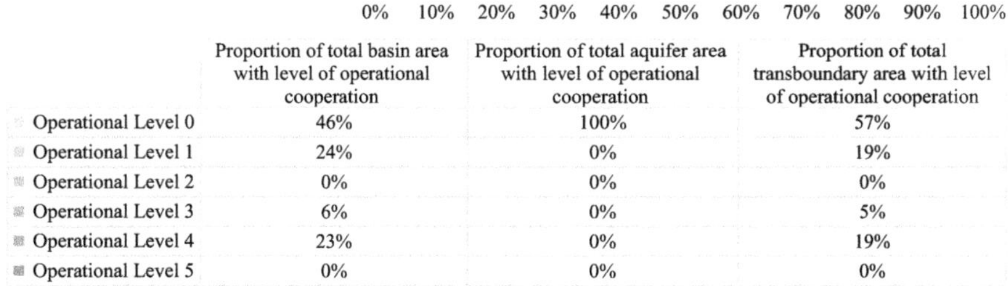

	Proportion of total basin area with level of operational cooperation	Proportion of total aquifer area with level of operational cooperation	Proportion of total transboundary area with level of operational cooperation
Operational Level 0	46%	100%	57%
Operational Level 1	24%	0%	19%
Operational Level 2	0%	0%	0%
Operational Level 3	6%	0%	5%
Operational Level 4	23%	0%	19%
Operational Level 5	0%	0%	0%

Figure 3.5 Results of Method 2 for Honduras.

Source: Reproduced with permission from *Measuring Transboundary Water Cooperation: Options for Sustainable Development Goal Target 6.5* (Global Water Partnership, 2017) written by Melissa McCracken for the Global Water Partnership (GWP) for its Technical Committee Background Papers series.

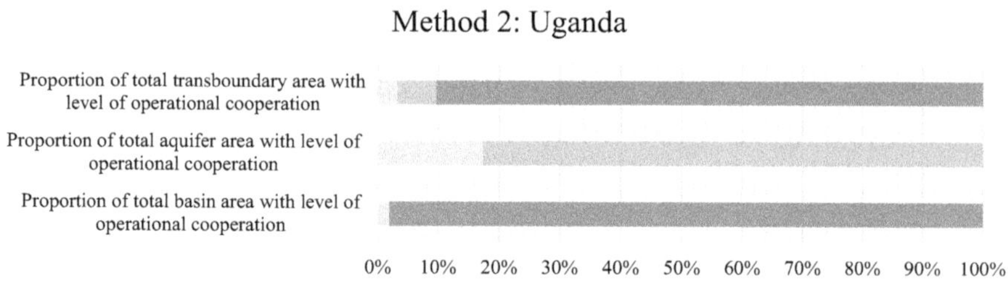

	Proportion of total basin area with level of operational cooperation	Proportion of total aquifer area with level of operational cooperation	Proportion of total transboundary area with level of operational cooperation
Operational Level 0	2%	17%	3%
Operational Level 1	0%	0%	0%
Operational Level 2	0%	83%	7%
Operational Level 3	0%	0%	0%
Operational Level 4	0%	0%	0%
Operational Level 5	98%	0%	90%

Figure 3.6 Results of Method 2 for Uganda.

Source: Reproduced with permission from *Measuring Transboundary Water Cooperation: Options for Sustainable Development Goal Target 6.5* (Global Water Partnership, 2017) written by Melissa McCracken for the Global Water Partnership (GWP) for its Technical Committee Background Papers series.

were excluded in Method 1. For example, as discussed above, the Ganges-Brahmaputra-Meghna River Basin in Bangladesh did not have operational cooperation in Method 1. However, when evaluating the criteria in Method 2, the area has cooperative efforts at Operational Level 4 – meaning 54% of Bangladesh's transboundary area has some degree of cooperation– when compared to 0% in Method 1, this is a significant increase. Upon further inspection of the results for Bangladesh, it is apparent that only the waters shared with Myanmar do not have some level of cooperation. The Lempa River Basin shows a similar situation for Honduras, as it also has Operational Level 4 for its cooperative efforts.

The level of cooperation better tracks and demonstrates the transboundary cooperation that is occurring in shared aquifers. In Method 1, all transboundary aquifer area was found not to have an operational cooperative arrangement. This result is expected as aquifers present more difficulties to cooperation given their inherent complexities, plus there are few examples of cooperation over shared aquifers, particularly when compared to the number of cooperative agreements that have been developed for shared surface water. However, by acknowledging smaller levels of cooperation, more cooperation over shared groundwater may be occurring than generally thought. For example, in Uganda, 83% of transboundary aquifer area has cooperative efforts at Operational Level 2. The two aquifers that contribute to this number underlie the Nile River; the *Cooperative Framework Agreement* specifically mentions groundwater, and some projects assessing groundwater done by the NBI with donor agencies have collected and shared data.

Having levels of cooperation, rather than a strict definition for operational cooperation, will give more visibility to the progress countries make toward transboundary cooperation in the international arena. Monitoring this will track progress and support the efforts countries make. Establishing a cooperative arrangement is a slow process – drafting and signing an agreement may take decades – therefore, a more nuanced means to monitor progress may provide more encouragement to continue the process.

Furthermore, by not requiring the fulfillment of all five criteria to be operational, Method 2 allows for some degree of acknowledgment of alternative cooperative efforts. The Goascorán River Basin, as discussed in Method 1, is recognized as having an Operational Level 3 cooperative effort despite lacking an agreement between El Salvador and Honduras, which excludes it as an operational cooperative arrangement in Method 1.

This method still has the same issues as Method 1 in that the definition of *arrangement*, the definition of *operational*, and the criteria that must be satisfied are procedural and normative. There is a greater degree of flexibility in how progress is monitored and reported, but the definitions still present a constrained view on what is appropriate for transboundary cooperation given the context, place, and sociopolitical situation. What is lacking from both Method 1 and Method 2 is the inclusion of substantive rules as a criterion for measuring transboundary cooperation. Substantive rules establish the rights and obligations for states sharing waters, which could also apply to other actors depending on the scale and context of cooperative efforts. Perhaps customary laws, such as 'reasonable and equitable use' or 'no significant harm,' should be included as criteria for operationality. Can the methodology for SDG Indicator 6.5.2 ensure that the transboundary cooperation that is being measured is effective if only procedural components are considered? This idea led to the development of Method 3. A reframing of operational cooperative arrangements away from inputs and toward substantive elements and the outputs of cooperation may present an alternative for measuring effective transboundary cooperation over shared waters.

Method 3: Typology of cooperation promoting effective water management

The third methodology for measuring transboundary water cooperation approaches the calculation from a different perspective than Methods 1 and 2. The current indicator (Method 1) and Method 2 define an arrangement as operational based on the process of cooperation. While the regular meeting of parties, exchange of information, joint management plans, and river basin organizations are firmly based on international law, they are strictly procedural in nature. Good process, however, does not automatically result in cooperation and effective management of shared waters. Therefore, this method aims to evaluate transboundary cooperation through a substantive lens over a procedural perspective. The goal is to present an alternative understanding of how transboundary cooperation can be conceptualized and evaluated based on the outcomes of a cooperative effort, rather than strictly on the process of cooperation – or an emphasis on output cooperation over input cooperation (Tarlock 2015). Furthermore, this perspective acknowledges that process is often context-specific; each place, basin, aquifer, country, and so on has a unique situation in which to develop integrated management that enhances cooperation and effectively manages shared waters. Therefore, this method has an alternative definition for cooperative arrangement and what constitutes an operational arrangement. A *cooperative arrangement*, in Method 3, is the cumulation of the cooperative efforts between riparians on a transboundary basin or aquifer. This includes formal agreements, organizations and other institutions, joint projects and plans, informal agreements, and organizations, as well as efforts by non-state actors. The systems of cooperative efforts are then evaluated to determine if they are operational.

Method 3 defines *operational* as substantive cooperation that results in effective water management or an increase in water security. *Operational* is not determined by a set of criteria that must be met but rather as a measure of the type of cooperation (defined below) that occurs within the country basin or aquifer area, along with the benefits of that cooperation. To calculate this measure, the type of operational cooperation is identified for each BCU or ACU. While the types of cooperation imply steps, countries do not need to progress through all the steps to achieve the final type of cooperation. Further, depending on the context, one type of cooperation may be the most appropriate, whereas the final type – Continuing Comprehensive Cooperation – may not. For example, in a BCU that composes very little of the total area of the basin and contributes very little to nothing in terms of volume of water to the catchment, Non-Cooperation or Preliminary Cooperation may be a more appropriate type of cooperation for that BCU than Continuing Comprehensive Cooperation. The types of cooperation and the focus on outcome cooperation are adapted from the GWP TEC Background Paper *Promoting Effective Water Management Cooperation among Riparians* by Tarlock (2015) and are as follows:

- *Non-Cooperation*: This type of cooperation is no cooperation. There is no formal or informal cooperative arrangement between the riparians, including agreements, river basin organizations, or dialogues.
- *Preliminary Cooperation*: Within this type of cooperation, riparians have expressed the intent to cooperate; this cooperation, however, has not been defined either substantively or procedurally. Riparian states can, for example, sign an agreement to develop transboundary cooperation or an RBO in the future. The signing of an agreement is not required to be in this type of cooperation, and countries can verbally express intent to cooperate or be entering negotiations.

- *Issue Cooperation*: A cooperative arrangement exists between riparian countries to address a specific issue or issues; management or governance issues are addressed in isolation with limited efforts toward shared benefits between the riparians or other actors. Alternatively, a state could act unilaterally to prevent harm or resolve conflict with another riparian. This could include the development of a joint agreement or commission to construct infrastructure or manage floods. Issues are categorized according to areas of primary interest initially developed by Hamner and Wolf (1997) and since adapted by Wolf et al. (2003), Giordano et al. (2013), and through this paper. These issues are: water quality, water quantity, aquatic ecosystem, hydropower, navigation, fishing, flood control/relief, economic development, joint management, irrigation, infrastructure/development, technical cooperation/assistance, border issues, and territorial issues (Hamner and Wolf 1997; Wolf, Yoffe, and Giordano 2003; Giordano et al. 2013).
- *Emerging Comprehensive Cooperation*: Riparian countries are developing or have recently developed a cooperative arrangement that establishes a legal framework for the shared management of the basin. Informal processes may also exist. Not all relevant[13] riparians may yet be included in the arrangement. The intention behind the cooperative efforts is to create shared benefits – these may have not yet come to fruition. The cooperative arrangement addresses multiple issues[14] such as quantity, quality, and so on, and includes coordination mechanisms. The issues included are appropriate for the basin and transboundary management; everything related to water management does not need to be included at the international level for there to be Emerging Comprehensive Cooperation. Collaboration at this level between riparians is more recent; outcomes of cooperation may not yet be realized. There may not be a history of cooperation with the country of interest and surrounding riparians.
- *Continuing Comprehensive Cooperation*: This type of cooperation is similar to Emerging Comprehensive Cooperation. Riparian countries have developed a cooperative arrangement for an ongoing legal framework for shared management of the basin. Informal (nongovernmental) coordination may also be in place to fill gaps or supplement formal institutions. All relevant riparians are included in the cooperative efforts, although multilateral arrangements are not required, provided there is basin-wide coverage through alternative cooperative efforts. States have continued to collaborate to address multiple issues related to the shared waters, and solutions include shared benefits. The issues included are appropriate for the basin; not all issues must be addressed at the international level for there to be Continuing Comprehensive Cooperation. Outcomes of the cooperative effort are apparent. The arrangement tracks with customary law, and riparians may be signatories of international conventions or have addressed the framework of international law in elements of their cooperative arrangements. Continuing Comprehensive Cooperation is different from Emerging Comprehensive Cooperation in that it has been actively

> occurring for at least a decade; this will be identified through positive interactions that have occurred between the country of interest and its riparians. In addition, it is different from Emerging Comprehensive Cooperation through the extent of basin coverage and inclusion of relevant riparians, as well as what issues and mechanisms are addressed by the cooperative arrangement.

This method does not meet the needs of the UN Statistics department, as it is based on qualitative data and has categorical results for the indicator, rather than numerical. However, the intent behind the development of this method is to highlight the deficits of the procedural-based method of SDG Indicator 6.5.2. Also, it identifies the variability needed to have successful cooperation, as each place and its shared waters are unique – governance should also be context-specific. Lastly, the method is developed to provide additional guidance, as a complement to the finalized SDG Indicator 6.5.2, to countries to help them gain an alternative perspective on the state of their transboundary cooperation. Basin managers have the institutional and local knowledge to place their cooperative efforts within the spectrum of cooperation. Using this method, they can further develop cooperative policies and actions that work toward reaching the type of cooperation that is most appropriate to their basin.

At its current level of development, this method is not feasible for global scale monitoring, as it requires an extensive understanding of the current transboundary situation. However, specific standardized metrics to measure the effectiveness of a cooperative arrangement for placement within a type of cooperation should be developed, which would improve the usability and versatility of the method.

The results of the survey data, additional literature, and international databases were used to determine which category of cooperation each ACU or BCU belonged in. The results of the calculation using Method 3 are displayed spatially in maps of each country – Figures 3.7 to 3.9 – and are accompanied by a table detailing the results – Tables 3.2 to 3.4; the results are examined in the following section.

What can we learn from comparing these three interpretations of transboundary water cooperation?

Tables 3.2 to 3.4 present the Method 3 tabular results for Bangladesh, Honduras, and Uganda, respectively. Overall, it is apparent that there are similar trends to those seen in the results of Method 2 (Figures 3.4 to 3.6). Both Bangladesh and Honduras have around half of their transboundary land area in Non-Cooperation. While this result may seem to place both states in a similar extent of cooperation, it is important to recognize the impact of these transboundary bodies' area in the calculation compared to the number of basins and aquifers. The majority of Bangladesh is within the Ganges-Brahmaputra-Meghna River Basin and the underlying East Ganges River Plain Aquifer; the other basins make

Table 3.2 Tabular results for Bangladesh with Method 3

Bangladesh			
Types of cooperation	*Proportion of total basin area (with Teesta River in total) (%)*	*Proportion of total aquifer area (%)*	*Proportion of total transboundary area (with Teesta River in total) (%)*
Non-Cooperation	0	100	45
Preliminary Cooperation	11	0	6
Issue Cooperation	88	0	49
Emerging Comprehensive Cooperation	0	0	0
Continuing Comprehensive Cooperation	0	0	0

Source: Reproduced with permission from *Measuring Transboundary Water Cooperation: Options for Sustainable Development Goal Target 6.5* (Global Water Partnership, 2017) written by Melissa McCracken for the Global Water Partnership (GWP) for its Technical Committee Background Papers series.

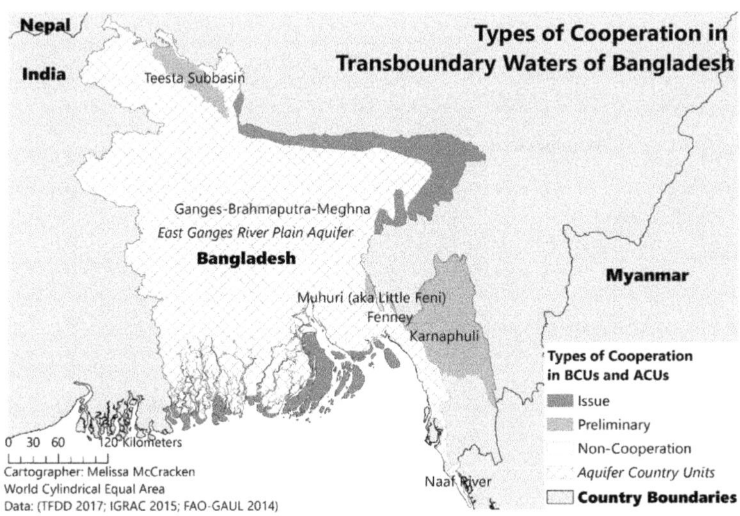

Figure 3.7 Types of cooperation in basin country units and aquifer country units in Bangladesh.

Source: Reproduced with permission from *Measuring Transboundary Water Cooperation: Options for Sustainable Development Goal Target 6.5* (Global Water Partnership, 2017) written by Melissa McCracken for the Global Water Partnership (GWP) for its Technical Committee Background Papers series.

Table 3.3 Tabular results for Honduras for Method 3

Honduras			
Types of cooperation	*Proportion of total basin area (%)*	*Proportion of total aquifer area (%)*	*Proportion of total transboundary area (%)*
Non-Cooperation	46	100	57
Preliminary Cooperation	24	0	19
Issue Cooperation	0	0	0
Emerging Comprehensive Cooperation	30	0	24
Continuing Comprehensive Cooperation	0	0	0

Source: Reproduced with permission from *Measuring Transboundary Water Cooperation: Options for Sustainable Development Goal Target 6.5* (Global Water Partnership, 2017) written by Melissa McCracken for the Global Water Partnership (GWP) for its Technical Committee Background Papers series.

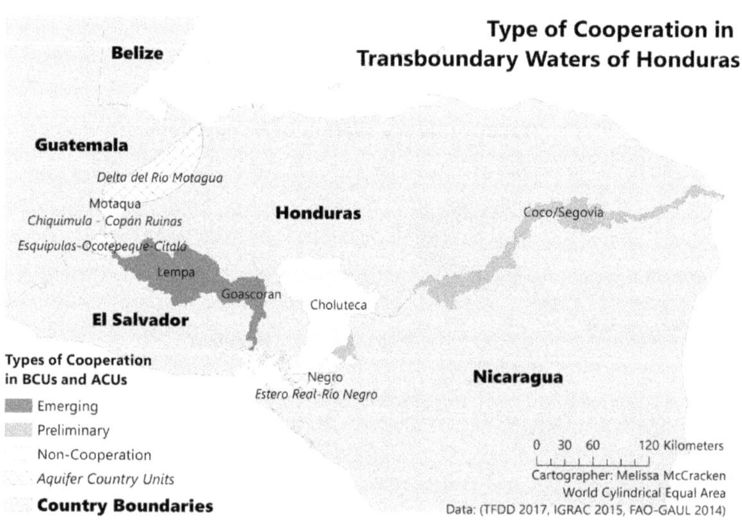

Figure 3.8 Types of cooperation in basin country units and aquifer country units in Honduras.

Source: Reproduced with permission from *Measuring Transboundary Water Cooperation: Options for Sustainable Development Goal Target 6.5* (Global Water Partnership, 2017) written by Melissa McCracken for the Global Water Partnership (GWP) for its Technical Committee Background Papers series.

Table 3.4 Tabular results for Uganda for Method 3

Uganda			
Types of cooperation	*Proportion of total basin area (with Lake Victoria in total) (%)*	*Proportion of total aquifer area (%)*	*Proportion of total transboundary area (with Lake Victoria in total) (%)*
Non-Cooperation	1	17	2
Preliminary Cooperation	0	83	5
Issue Cooperation	0	0	0
Emerging Comprehensive Cooperation	77	0	72
Continuing Comprehensive Cooperation	21	0	20

Source: Reproduced with permission from *Measuring Transboundary Water Cooperation: Options for Sustainable Development Goal Target 6.5* (Global Water Partnership, 2017) written by Melissa McCracken for the Global Water Partnership (GWP) for its Technical Committee Background Papers series.

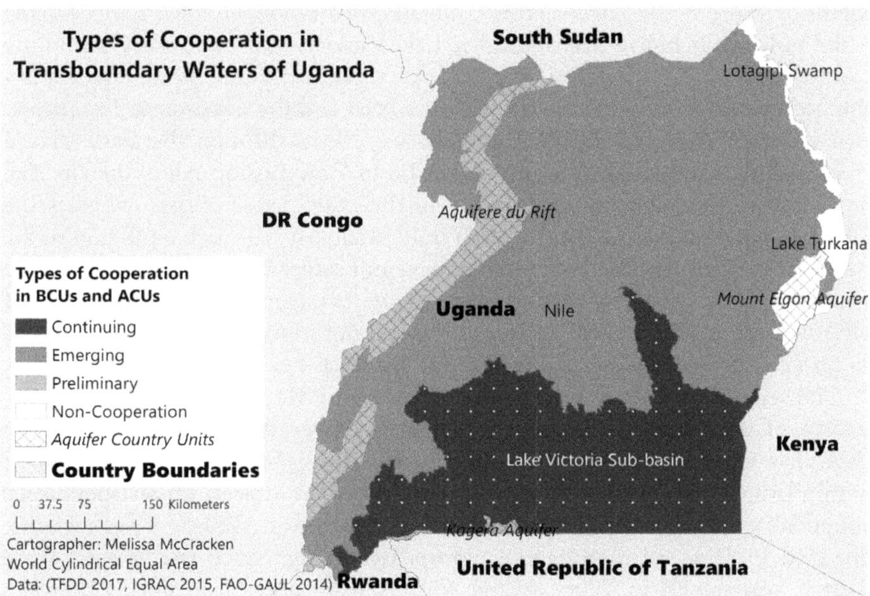

Figure 3.9 Types of cooperation in basin country units and aquifer country units in Uganda.

Source: Reproduced with permission from *Measuring Transboundary Water Cooperation: Options for Sustainable Development Goal Target 6.5* (Global Water Partnership, 2017) written by Melissa McCracken for the Global Water Partnership (GWP) for its Technical Committee Background Papers series.

up only about 5% of the transboundary area. The Non-Cooperation in the East Ganges Plain Aquifer overshadows the cooperation that is occurring in all but one of the transboundary rivers (excluding the Ganges-Brahmaputra-Meghna) due to the magnitude of the area that weights the calculation. Honduras, contrastingly, has several basins and aquifers of similar areas; half of the river basins have Non-Cooperation and half have some type of cooperation occurring. In the SDG Indicator calculations, area weights the contributions of particularly large basins or aquifers heavily. When considering the values of the SDG Indicator 6.5.2, users should recognize this limitation, as the area of a transboundary basin or aquifer does not necessarily correlate with the importance or priority for transboundary cooperation. Alternative methods for the indicator have been proposed that do not consider area but instead look at the proportion of countries that are riparian and have an agreement (AbuZeid 2016).

When comparing the results in the above tables to the results for Method 2, there are similarities; however, there are several differences worth highlighting and discussing. First, the Nile Basin in Uganda was found to have operational cooperation with Method 1, Operational Level 5 cooperation in Method 2, and Emerging Comprehensive Cooperation in Method 3. These results are relatively consistent across the methods; the notable difference is how the Lake Victoria Sub-basin (Figure 3.9) compares to the greater Nile Basin. In Methods 1 and 2, the Lake Victoria Sub-basin would receive the same scoring as the Nile Basin, but in Method 3, the Lake Victoria Sub-basin has Continuing Comprehensive Cooperation. The Lake Victoria Sub-basin, while also applicable to the cooperative efforts through the NBI and the *Cooperative Frameworks Agreement* (CFA), has additional cooperative efforts through the East African Community and the Lake Victoria River Basin Commission. All of the riparian states within the sub-basin participate in the cooperative efforts, whereas the DRC, Egypt, Sudan, and South Sudan only participate through a portion of the cooperative efforts in the Nile, as they have not ratified the CFA. Furthermore, while the NBI – and its predecessors – have been operating since 1999, the CFA was not reached until 2010, and it is still not in force, whereas the *Protocol for Sustainable Development of Lake Victoria Basin* has been ratified since 2003.

The second difference between Method 2 and Method 3 is in the categorization of the Ganges-Brahmaputra-Meghna River Basin. In Method 2, the BCU has Operational Level 4 cooperation, and in Method 3, the basin falls within Issue Cooperation. The BCU met most of the criteria for an operational cooperative arrangement in Methods 1 and 2; however, Method 3 takes a more nuanced look at the cooperative efforts occurring. Since the *Ganges Water Treaty*[15] was signed in 1996, several MOUs were previously signed between India and Bangladesh over the Ganges, and the Indo-Bangladesh JRC has been operating since 1972, cooperation could appear to be Continuing. However, the other riparians to the basin have not engaged with any other cooperative efforts with Bangladesh over the shared waters; Nepal, India, and Bangladesh have broached collaborating to increase the storage potential and augment dry season flows, but the political situation prohibits any further collaborative efforts (Dhungel 2013). In addition, the cooperative efforts between India and

Bangladesh are issue-specific: the *Ganges Water Treaty* and the Joint Committee handle water quantity issues while the JRC deals mainly with flood control/relief and technical cooperation/assistance. Issues relating to joint management appear to be dealt with in a 'siloed' fashion rather than through integration, which could potentially present unrealized shared benefits.

The final key difference to note is with the Goascorán Basin in Honduras. In Methods 1 and 2, the basin had no operational cooperation and Operational Level 3 cooperative arrangements, respectively. As discussed in the Method 1 and Method 2 sections, the emphasis on procedural criteria overlooked the cooperative efforts ongoing in the basin. Method 3 places the basin within Emerging Comprehensive Cooperation. While national actors are not actively participating in the joint management of the basin and there is no formal agreement between the two riparians, regional and local actors, as well as members of the public and private sector with the assistance of the IUCN Bridge program, have developed an institution for cooperation in the basin. While this effort is newly restarted in 2011, recognizing its efforts as Emerging Comprehensive Cooperation could incentivize international donors to continue to support the process or place pressure on the states to become involved. Ultimately, these examples illustrate three key points: (1) Method 3 considers the basin-wide cooperative efforts and the coverage of transboundary cooperation when categorizing the type of cooperation that is occurring within a particular BCU or ACU, (2) creating lasting cooperative efforts resulting in positive outcomes that continue to be effective is equally – if not arguably – more important than having all the 'key aspects' comprising transboundary cooperation, and (3) the type of cooperation established is dependent on the political will of the riparian states and the context within which the waters are shared.

The above discussion focuses on differences arising from a substantive lens to define operational versus a procedural lens. Method 3, however, also adapted the definition of a cooperative arrangement. This method broadens the conceptualization of arrangement to include the cumulation of all cooperative efforts and evaluates their operationality based on the combination. Non-state actors often play an important role in transboundary cooperation (Giordano et al. 2013); this broadening of the definition acknowledges their contribution. Two basins – the Lempa River Basin and the Coco/Segovia River Basin – in Honduras have nontraditional actors and alternative cooperative arrangements that contribute to the shared management of the rivers. The Lempa River has a state-to-state agreement as well as a national level joint management body. This arrangement, the Tri-National Commission and the Trifinio Plan, emphasizes sustainable development in the Trifinio Region, which includes the Upper Lempa River Basin. Historically, the focus has not been on the joint management of shared waters, and previously only indirect benefits to the shared water occurred – it was not until recently that a shift toward integrated water resources management occurred at this national scale. To fill the gaps, 26 municipalities plus actors from public and private sectors established the *Mancommunidad Trinacional Fronteriza Río Lempa* to address development problems in the region, the lack of public policy from the national level, and the lack of public participation. The

Mancommunidad has established an *Aguas Compartidas* program to address these issues as they are specifically related to the Upper Lempa Basin.

Similar to the Upper Lempa Basin, the Coco/Segovia River Basin has alternative cooperative arrangements and non-state actors. The Coco/Segovia River Basin, shared between Honduras and Nicaragua, has no national cooperative actions occurring to jointly manage the shared waters. Supported by UNEP, UNOPS lead the program Supporting Sustainable Water Management in the Coco/Segovia River Basin; this program aimed to strengthen local capacity for integrated water management originating at the micro-basin level and between the departments of Nueva Segovia and El Paraíso in Nicaragua and Honduras, respectively. The project concluded in 2012 and resulted in management plans for micro- and sub-watersheds in the Coco/Segovia River Basin. The situation in these two basins reflects the need to strengthen the definition of what constitutes a cooperative arrangement for SDG Indicator 6.5.2. Non-state actors and informal institutions that do not fit with the current conceptualization of *arrangement* play a significant role, particularly when there is a lack of support or resources at the national level for transboundary cooperation.

As mentioned in Method 3 explanation, this method's typology of cooperation implies a stepwise path culminating in Continuing Comprehensive Cooperation that has lasting cooperative outcomes. However, despite this implication, having Continuing Comprehensive Cooperation may not be necessary for a particular basin or aquifer depending on the context, nor do countries need to progress through every stage. The Lake Turkana Basin, Figure 3.9, shared between Ethiopia, Kenya, South Sudan, and Uganda, is an endorheic lake. The lake and most of its tributaries are in Ethiopia and Kenya; the BCU in Uganda contributes very minimally to the basin – both in terms of catchment area and volumetric flow. Ethiopia and Kenya have begun to discuss the creation of a bilateral agreement to manage the basin jointly. Given Uganda's minimal contribution to the basin, having Non-Cooperation may be an appropriate type of cooperation for its BCU, which may boost cooperation between the major basin states. Alternatively, Preliminary Cooperation could occur within Uganda's BCU by stating the country would be willing to cooperate over the basin in the future, if needed, but will allow Ethiopia and Kenya to manage the lake jointly. Essentially, the transboundary cooperation occurring is operational, whether or not it meets procedural or substantive criteria as defined by international law. If cooperative efforts are functioning within the political constructs and the context of the basin or aquifer and resulting in effective positive outcomes as defined in that context, then why should it not be considered operational?

There are limitations to this method for measuring transboundary cooperation. Placement of a BCU or ACU within a type of cooperation is currently somewhat subjective and requires in-depth knowledge of the related cooperative efforts. Therefore, it is not feasible to conduct a global scale study, particularly at the data collection and monitoring scale required of the SDGs. Objective criteria that avoid emphasizing a normative framework need to be developed for this method. The role and purpose of this method are to illustrate alternative means to evaluate the operationality of cooperative arrangements and what actually comprises a cooperative arrangement. The goal is to create a tool for basin managers and practitioners to evaluate transboundary cooperation without restricting how cooperative efforts are established or how those efforts function. The typology of cooperation attempts to reflect the variety of cooperative efforts that can occur without placing too much emphasis on what is input. Additionally, it attempts to create space to develop alternative or pragmatic cooperative solutions that are place-based and within the sociopolitical situation.

Conclusion

This chapter evaluates three methodologies for calculating the indicator in order to aid in maximizing the benefit from SDG Indicator 6.5.2 and demonstrating the two main challenges this book is addressing – defining cooperation and what makes it effective. Method 1 is the official method for SDG Indicator 6.5.2 and was promoted to Tier I in November 2018 (IAEG-SDG 2019). It defines a cooperative arrangement as operational if it meets all of the procedural criteria; these criteria are drawn from principles in the universal water conventions. Method 2 is founded in the same conceptualization as Method 1. To determine if an arrangement is operational, this method does not require that all criteria must be fulfilled. As long as one criterion is met, the area is considered to have some form of operational cooperation. This creates levels of operational cooperation. Method 3 takes an alternative perspective based on substantive principles and focuses on the outcomes of effective cooperation, rather than procedural criteria. Operational, in Method 3, is based on substantive cooperation, which is determined by a typology of cooperation based on categories developed in *Promoting Effective Water Management Cooperation among Riparian Nations* (Tarlock 2015). Table 3.5 summarizes the strengths and weaknesses of the three methodologies for evaluating operational arrangements for transboundary water cooperation at the national level.

From this analysis, we can draw two sets of conclusions related to the two goals of the chapter. The first order of conclusions is directly related to the analysis of the three methodologies and maximizes the benefit from SDG Indicator 6.5.2. There are several key points to summarize from the evaluation of SDG Indicator 6.5.2 presented in this chapter. First, it is necessary to have clear and consistent definitions of what is a transboundary basin, a transboundary aquifer, an arrangement, and what makes an arrangement operational. Defining arrangement and operational are of particular importance for this indicator. The current definition of arrangement attempts to build flexibility but could lead to confusion when identifying an arrangement. Operational is used as the variable for comparison between methods, and through this, it is apparent that how operational is defined can significantly impact what cooperative arrangements – or efforts – are recognized by SDG Indicator 6.5.2. The definition of operational, used in Method 1 and adapted in Method 2, is based on procedural and normative criteria, which may

Table 3.5 Summary of strengths and weaknesses of the three measurement methods for *operational* arrangements for transboundary water cooperation, including SDG Indicator 6.5.2

SDG Indicator 6.5.2 Methodologies	*Strengths*	*Weaknesses*
Method 1	• Operational criteria are based on water conventions • Criteria incorporate the value in countries having an agreement and joint body • Relatively straightforward to determine if operational • Results in a single numerical value that meets needs for global monitoring • Results have the potential to be aggregated and disaggregated if criteria satisfaction are shared	• *Arrangement* has an inconsistent definition that could lead to confusion • Masks cooperation by requiring all criteria be fulfilled - binary • Criteria for operationality are procedural and normative • Spatial data for basins and aquifers may be hard to access or develop, particularly for areas lacking studies of hydrogeology • Survey data collection for operational criteria and arrangements has the potential to be inconsistent between countries and may be limited
Method 2	• Operational criteria are based on water conventions • Criteria incorporate the value in countries having an agreement and joint body • Levels of Operational Cooperation recognize a greater extent of transboundary cooperation occurring • Allows for some flexibility in how basins develop their cooperative arrangements • Creates incentives for developing transboundary cooperation by tracking progress more explicitly	• *Arrangement* has an inconsistent definition that could lead to confusion • Criteria for operationality are procedural and normative • Spatial data for basins and aquifers may be hard to access or develop, particularly for areas lacking studies of hydrogeology • Survey data collection for operational criteria and arrangements has the potential to be inconsistent between countries and may be limited • Does not present a single value for each country, which does not meet the needs for global SDG monitoring
Method 3	• Allows for flexibility in cooperative efforts • Recognizes cooperation is dependent on the political will and social-political context of the shared water • Emphasizes cooperative efforts that result in continued positive outcomes • Uses a widened definition of cooperative arrangement to include the cumulation of cooperative efforts • Acknowledges the role of non-state and local actors in cooperative efforts	• Does not present a single value for each country, which does not meet the needs for global SDG monitoring • Spatial data for basins and aquifers may be hard to access or develop, particularly for areas lacking studies of hydrogeology • Degree of subjectivity in assigning Type of Cooperation, objective criteria are needed pending further research • Requires in-depth knowledge of a variety of scales of cooperative efforts occurring

exclude alternative cooperative arrangements. Further, the binary categorization in Method 1 has the potential to mask ongoing cooperation that does not meet all the criteria but is an effective means of cooperation. Furthermore, a prescriptive framework for transboundary cooperation has the potential to limit adaptability in forming cooperative arrangements that reflect the context of the basin. Cooperation between riparians is mostly determined by the political will of the riparian nations (Wouters 2013; Zeitoun and Mirumachi 2008). For example, the sociopolitical context may not allow for the signing of an agreement or information exchange. Therefore, water governance and transboundary cooperation over shared waters should allow for place-based specifics and cooperative efforts that fit the sociopolitical environment (Giordano and Shah 2014; Jensen 2013). Similarly, measurement and monitoring of transboundary cooperation should equally reflect the diversity of cooperative efforts. Non-state actors, municipalities, and public and private sectors, for example, may develop cooperative efforts over shared waters, particularly if there is a limited response from national governments. These efforts can result in effective outcomes of cooperation. Method 3's typology of cooperation placed a greater emphasis on the substantive elements. While currently not feasible for a global study at the scale of the SDGs, this method aims to be a tool for basin managers, academics, and shared water users to evaluate transboundary cooperation and provide an alternative viewpoint for measuring progress toward transboundary cooperation.

The second order of conclusions relates to the two broader goals of this book – defining cooperation and determining what makes it effective. The importance of SDG Indicator 6.5.2 illustrates two factors. First that there is growing global consensus that transboundary water cooperation is a means to encourage sustainable development and reduce conflict between states over shared waters. Second that there is a demand for measuring and monitoring cooperation. From the comparison and evaluation of the three methods for evaluating operational arrangements for transboundary water cooperation in the manner of SDG Indicator 6.5.2, this chapter demonstrates that even minor differences in how transboundary water cooperation is defined can have a dramatic impact on how it is depicted, measured, and understood. Nevertheless, when we consider how many definitions of transboundary water cooperation exist in the international water field and their breadth (see a sample of definitions in Chapter 2), the scope and potential impacts of this challenge of a lack of a clear, consistent definition are readily apparent. The lack of a commonly understood and utilized definition can limit growth and understanding of cooperation and inhibit the development and sharing of knowledge as to what cooperation is and what makes it effective. This latter question is ultimately the aim of many frameworks and definitions of cooperation – what aspects make cooperation have positive outcomes and be effective? – which is the second challenge of this book. As demonstrated by comparing Methods 1 and 2 with Method 3, a good process does not automatically result in cooperation and effective management of shared waters. This chapter has illustrated the two challenges at the core of this book. In the following chapters, Part II addresses the question: how do we define transboundary water coordination in a clear, consistent manner, while Part III addresses what needs to be in place for there to be effective cooperation.

Notes

1 Chapter 3 is reproduced (with revisions and updates) from Measuring transboundary water cooperation: options for Sustainable Development Goal Target 6.5 (Global Water Partnership, 2017) written by Melissa McCracken for the Global Water Partnership (GWP) for its Technical Committee Background Papers series.

2 See Sindico 2016 for full discussion on links between SDG Indicator 6.5.2, SDG 6, and other SDGs.

3 MDG Target 7.C: Halve, by 2015 the proportion of the population without sustainable access to safe drinking water and basic sanitation (United Nations n.d.).

4 UNECE Convention on the Protection and Use of Transboundary Watercourses and International Lakes (Water Convention) was amended in Feb 2013 to become a global framework. As of 1 March 2016, countries outside of the ECE regions can accede to the convention (UNECE 2016).

5 This book follows the naming convention of the Transboundary Freshwater Dispute Database for the Ganges-Brahmaputra-Meghna River Basin. This is based on the definition used for an international river basin which is determined by a common terminus.

6 SDG Indicator 6.5.2 was elevated to a Tier I indicator in November 2018 (IAEG-SDG 2019).

7 The chapter method deviates from the methodology for Indicator 6.5.2, in that it uses global datasets for the delineations of transboundary river basin and aquifers. This is to allow comparison between the results of the three methods discussed in this paper. For more detail, the methodology document can be found here: www.unwater.org/publications/step-step-methodology-monitoring-transboundary-cooperation-6-5-2/.

8 The template for the questionnaire was presented at the Eleventh Meeting of the Working Group on Integrated Water Resources Management on 18–19 October 2016 in Geneva. It can be found here: www.unece.org/fileadmin/DAM/env/documents/2016/wat/10Oct_18-19WGIWRM/WG.1_2016_INF5_reporting.pdf.

9 Survey responses were only received from Bangladesh and Honduras. Evaluation of Uganda's transboundary cooperation only used the results of secondary research and available literature.

10 The International Freshwater Treaties Database and the International River Organization Database are available as part of the Transboundary Freshwater Dispute Database at: https://transboundarywaters.science.oregonstate.edu/.

11 Consistency in responses to the questionnaires varied; therefore, additional research was used to supplement survey responses to determine if the cooperative arrangements met the required criteria.

12 The Ganges-Brahmaputra-Meghna BCU in Bangladesh has a cooperative arrangement, several joint institutions, and committees, and has some data and information exchange related to specific issues, but there is no coordinated management plan or joint objectives for the basin.

13 Relevant riparians are those basin states that contribute significant area or flow, have reasonable population size living in the basin area, or a substantial impact on the basin.
14 The same issue categories will be used as detailed in issue cooperation.
15 Full title: *Treaty between the government of the Republic of India and the government of the People's Republic of Bangladesh on sharing of the Ganga/Ganges waters at Farakka.*

References

AbuZeid, K. 2016. "Draft 3rd Arab State of the Water Report Indicators." Center for Environment & Development for the Arab Region & Europe (CEDARE), and Arab Water Council.

Ahmed, Abu Musa Md. Motaher, and Kingshuk Roy. 2007. "Utilization and Conservation of Water Resources in Bangladesh." *Journal of Developments in Sustainable Agriculture* 2 (1): 35–44.

Alexovich, Ariel. 2015. "UN Board: New Global Arena Needed to Confront Water Issues." *United Nations Sustainable Development* (blog). November 18, 2015. www.un.org/sustainabledevelopment/blog/2015/11/new-global-arena-needed-to-confront-water-challenges-un-board-says/.

AMCOW (African Ministers' Council on Water). 2013. "Tunis Outcome Document for the Water Sector Post-2015 Thematic Consultations." www.amcow-online.org/images/docs/outcomes_of_the_tunis_post_2015_water_consultations.pdf.

Baca, Pedro, José Iván García Olivera, Fernando Leiva, Rene Rigoberto Perez Carrasco, Domingo Rivas, Pastora Rivera Centeno, Laura Karina Rodríguez Sarantes, Alex David Zapata, and Nestor Lopeez Nolasco. 2012. *Planes Para El Manejo de La Cuenca Media y Alta Del Rio Coco*. Managua: UNOPS and SIMAS.

Baca, Pedro, José Iván García Olivera, Fernando Leiva Pérez Carrasco, Domingo Rivas, Pastora Rivera Centeno, Laura Karina Rodríguez Sarantes, and Alex David Zapata. 2012. *Como Elaborar Un Plan de Manejo En La Cuence Media y Alta Del Rio Coco*. Managua: UNOPS and SIMAS.

Bhuiyan, ABM Faruquzzaman, and M.M. Hossain. 2006. "Transboundary Issues and Uncertainty in Water Management in Bangladesh." In *World Environmental and Water Resource Congress*, 1–10. American Society of Civil Engineers. http://ascelibrary.org/doi/abs/10.1061/40856%28200%29249.

Buch, Mario Samuel, and Jose Mario Guevara. 2010. "Trifinio: Transboundary Aquifer Systems in the Upper Lempa River Basin, El Salvador, Guatemala and Honduras, in Central America." Presented at ISARM2010 Transboundary Aquifers: Challenges and New Directions on 6-8 December 2010. https://hispagua.cedex.es/sites/default/files/hispagua_documento/documentacion/documentos/acuifero_trifinio.pdf.

Dhungel, Dwarika. 2013. "Regional Cooperation on the Ganga Basin: Yet a Mirage?" *The Asia Foundation: Issue Brief*, September. https://asiafoundation.org/resources/pdfs/ORFIssuebrief61DwarikaN.Dhungelformail.pdf.

DWRM. 2011. "The Declining Trends of Water Resources in Uganda." Entebbe: Water Resources Monitoring and Assessment Division in the Directorate of Water Resources Management.

EAC. 2000. "Treaty Establishing the East African Community (EAC)." www.eac.int/treaty/.

FAO. 2014. "Bangladesh." FAO AQUASTAT. www.fao.org/nr/water/aquastat/countries_regions/BGD/BGD-CP_eng.pdf.

———. 2015a. "Honduras." FAO AQUASTAT. www.fao.org/nr/water/aquastat/countries_regions/HND/index.stm.
———. 2015b. "Uganda." FAO AQUASTAT. www.fao.org/nr/water/aquastat/countries_regions/UGA/print1.stm.
Fundación Hondureña de Ambiente y Desarrollo VIDA. 2008. "Transboundary: Trans-Border Management Group for the Conservation of the Environment of the Goascorán River, Honduras and El Salvador." Global Water Partnership – IWRM Toolbox. www.gwp.org/en/ToolBox/CASE-STUDIES/Americas--Caribbean/Transboundary-Trans-border-management-group-for-the-conservation-of-the-environment-of-the-Goascoran-River-Honduras-and-El-Salvador-320/.
Giordano, Mark, Alena Drieschova, James A. Duncan, Yoshiko Sayama, Lucia De Stefano, and Aaron T. Wolf. 2013. "A Review of the Evolution and State of Transboundary Freshwater Treaties." *International Environmental Agreements: Politics, Law and Economics* 14 (3): 245–64. https://doi.org/10.1007/s10784-013-9211-8.
Giordano, Mark, and Tushaar Shah. 2014. "From IWRM Back to Integrated Water Resources Management." *International Journal of Water Resources Development* 30 (3): 364–76. https://doi.org/10.1080/07900627.2013.851521.
Global Water Partnership. 2017. "Measuring Transboundary Water Cooperation: Options for Sustainable Development Goal Target 6.5." Written by Melissa McCracken for the TEC Background Paper 23. Stockholm: Global Water Partnership.
GWP. 2016. "Integrated Water Resources Management in Central America: The Challenge of Transboundary Water Management - Unpublished Draft." Technical Focus Paper. Stockholm: Global Water Partnership.
GWP Central America. 2015. *Situacion de las Recursos Hidricos en Centroamerica: Honduras.* Tegucigalpa, Honduras: Global Water Partnership.
Hamner, Jesse, and Aaron T. Wolf. 1997. "Water: I. Patterns in International Water Resource Treaties: The Transboundary Freshwater Dispute Database." *Colorado Journal of International Environmental Law and Policy* 1997: 157–219.
Hossain, Ishtiaq. 1998. "Bangladesh-India Relations: The Ganges Water-Sharing Treaty and Beyond." *Asian Affairs: An American Review* 25 (3): 131–50.
IAEG-SDG. 2016. "Introduction to Provisional Tiers of Global SDG Indicators." Presented at the 3rd Meeting of the Inter-agency and Expert Group on Sustainable Development Goal Indicators, Mexico City, April 30. http://unstats.un.org/sdgs/files/meetings/iaeg-sdgs-meeting-03/3rd-IAEG-SDGs-presentation-UNSD--Introduction-to-indicator-tiers.pdf.
———. 2019. "Tier Classification of Global SDG Indicators." UN Statistics. https://unstats.un.org/sdgs/files/Tier_Classification_of_SDG_Indicators_22_May_2019_web.pdf.
IGRAC, and UNESCO-IHP. 2015. "Transboundary Aquifers of the World Map 2015." International Groundwater Resources Assessment Centre. 2015. www.un-igrac.org/resource/transboundary-aquifers-world-map-2015.
Jensen, Kurt Mørck. 2013. "Swimming Against the Current: Questioning Development Policy and Practice." *Water Alternatives* 6 (2): 276.
Kasimbazi, Emmanuel B. 2010. "The Impact of Colonial Agreements on the Regulation of the Waters of the River Nile." *Water International* 35 (6): 718–32. https://doi.org/10.1080/02508060.2010.533642.
———. 2015. "Uganda: Sovereignty and International Water Agreements." In *A History of Water*, edited by Terje Tvedt, Owen McIntyre, and Tadesse Kassa Woldetsdik, Vol 2: 686–707. Series III. London: I.B.Tauris & Co. Ltd.

Kilimani, Nicholas. 2013. "Water Resource Accounts for Uganda: Use and Policy Relevancy." *Economic Research Southern Africa*, August. www.up.ac.za/media/shared/61/WP/working_paper_365.zp39427.pdf.

Leb, Christina. 2015. "One Step at a Time: International Law and the Duty to Cooperate in the Management of Shared Water Resources." *Water International* 40 (1): 21–32. https://doi.org/10.1080/02508060.2014.978972.

López, Alexander. 2004. *Environmental Conflicts and Regional Cooperation in the Lempa River Basin: The Role of Central America's Plan Trifinio*. Berlin: Adelphi Research.

MacQuarrie, Patrick, Rebecca Welling, Nazareth Porras, and Rocio Córdoba. 2013. *Goascorán Basin: Honduras and El Salvador*. Gland, Switerland: IUCN.

Mancomunidad Trinacional Fronteriza Rio Lempa. 2016. "Mancomunidad Trinacional Fronteriza Rio Lempa: Agua Compartidas." 2016. www.trinacionalriolempa.org/.

McCaffrey, Stephen C. 2003. "The Need for Flexibility in Freshwater Treaty Regimes." *Natural Resources Forum* 27 (2): 156–62. https://doi.org/10.1111/1477-8947.00050.

Medina, Lucile. 2014. "Gestion des bassins hydrographiques transfrontaliers et institution nalisation: l'initiative du Groupe de Gestion Binational du Bassin du Río Goascorán (Honduras-El Salvador)." *Regions & Cohesion* 4 (2): 54–80.

Mirchandani, Maya. 2016. "The Teesta Water Dispute: Geopolitics, Myth and Economics." *Observer Research Foundation*, September 21, 2016. www.orfonline.org/research/teesta-water-dispute/.

Nanni, Marcella. 2016. "Water Challenges in the IGAD Region: Towards New Legal Frameworks for Cooperation." *Water International* 41 (4): 635–51. https://doi.org/10.1080/02508060.2016.1169620.

Nishat, Ainun, and Islam M. Faisal. 2000. "An Assessment of the Institutional Mechanisms for Water Negotiations in the Ganges-Brahmaputra-Meghna System." *International Negotiation* 5 (2): 289–310.

Nsubuga, Francis N.W., Edith N. Namutebi, and Masoud Nsubuga-Ssenfuma. 2014. "Water Resources of Uganda: An Assessment and Review." *Journal of Water Resource and Protection* 06 (14): 1297–1315. https://doi.org/10.4236/jwarp.2014.614120.

Paisley, Richard, and Taylor W. Henshaw. 2013. "Transboundary Governance of the Nile River Basin: Past, Present and Future." *Environmental Development* 7 (July): 59–71. https://doi.org/10.1016/j.envdev.2013.05.003.

Rahaman, Muhammad Mizanur. 2006. "The Ganges Water Conflict: A Comparative Analysis of 1977 Agreement and 1996 Treaty." *Asteriskos* 1 (2): 195–208.

Republica de Honduras. 2009. "Decreto No. 181-2009: Ley General de Aguas." *La Gaceta Diario Oficial de La República de Honduras* No. 32,088: A1–A39.

Salman, Salman M.A. 2013. "The Nile Basin Cooperative Framework Agreement: A Peacefully Unfolding African Spring?" *Water International* 38 (1): 17–29. https://doi.org/10.1080/02508060.2013.744273.

Saruchera, D., and J. Lautze. 2015. "Measuring Transboundary Water Cooperation: Learning from the Past to Inform the Sustainable Development Goals." International Water Management Institute (IWMI). www.iwmi.cgiar.org/publications/iwmi-working-papers/iwmi-working-paper-168/.

Schweizerische Eidgenossenschaft. 2013. "Swiss Position Paper on Water in the Post-2015 Agenda." Schweizerische Eidgenossenschaft. www.swisswaterpartnership.ch/wp-content/uploads/2014/07/2013.12.20_Swiss_Position_on_Water_in_Post2015_Agenda.pdf.

Sindico, Francesco. 2016. *Transboundary Water Cooperation and the Sustainable Development Goals*. Paris: UNESCO-IHP.

Sood, Aditya., and Bala Krishna Prasad Mathukumalli. 2011. "Managing International River Basins: Reviewing India-Bangladesh Transboundary Water Issues." *International Journal of River Basin Management* 9 (1): 43–52.

Statute of the Indo-Bangladesh Joint Rivers Commission. 1972. Available at: www.fao.org/faolex/results/details/en/c/LEX-FAOC203668/

Suryanarayanan, Sowmya. 2010. "India and Bangladesh: Teesta River Agreement." Strategic Foresight Group. May 2010. www.strategicforesight.com/inner-articles.php?id=189#.Ue5Tlqz2v1U.

Tarlock, Dan. 2015. "Promoting Effective Water Management Cooperation Among Riparian Nations." *TEC Background Papers*, no. 21. www.gwp.org/Global/ToolBox/Publications/Background%20papers/GWP_TEC21_web.pdf.

Taylor, Paul. 2013. "National Stakeholder Consultations on Water: Supporting the Post-2015 Development Agenda." The Post-2015 Water Thematic Consultation. Stockholm: Global Water Partnership. https://sustainabledevelopment.un.org/content/documents/1815nationalstakeholder.pdf.

TFDD. 2021. "Transboundary Freshwater Dispute Database." College of Earth, Ocean and Atmospheric Sciences, Oregon State University. 2021. https://transboundarywaters.science.oregonstate.edu/.

UN Statistics. 2016a. "Metadata Goal 6 Ensure Availability and Sustainable Management of Water and Sanitation for All." United Nations. http://unstats.un.org/sdgs/files/metadata-compilation/Metadata-Goal-6.pdf.

———. 2016b. "Work Plans for Tier III Indicators." United Nations.

UNECE. 2016. "Water Convention." United Nations Economic Commission for Europe. 2016. www.unece.org/env/water.

UNECE Water Secretariat. 2020. "Step-by-Step Monitoring Methodology for SDG Indicator 6.5.2 2020." UN-Water. www.unwater.org/publications/step-step-methodology-monitoring-transboundary-cooperation-6-5-2/.

UNGC (United Nations Global Compact). 2013. "Corporate Sustainability and the United Nations Post-2015 Development Agenda." Report to the United Nations Secretary-General. United Nations Global Compact. www.stakeholderforum.org/fileadmin/files/UNGC_Post2015_Report.pdf.

United Nations. n.d. "United Nations Millennium Development Goals." United Nations. Accessed November 1, 2016. www.un.org/millenniumgoals/environ.shtml.

United Nations Environment Programme, ed. 2007. *Hydropolitical Vulnerability and Resilience along International Waters: Latin America and the Caribbean*. Nairobi: United Nations Environment Programme.

United Nations Environment Programme and United Nations Development Programme . 2012. *Status Report on the Application of Integrated Approaches to Water Resources Management*. Nairobi, Kenya: United Nations Environment Programme.

United Nations University and United Nations Office for Sustainable Development. 2013. *Catalyzing Water for Sustainable Development and Growth: Framing Water within the Post 2015 Development Agenda: Options and Considerations: Report*. United Nations University – Institute for Water, Environment, and Health, UN Office of Sustainable Development Stockholm Environment Institute. http://site.ebrary.com/id/10863310.

UNOPS. 2012. "Supporting Sustainable Water Management in the Coco River Basin." UNOPS. www.unops.org/SiteCollectionDocuments/Factsheets/English/Success%20Stories/GBL_PJFS_COCO_EN_0211.pdf.

UN-Water. 2013. *Water Security and the Global Water Agenda: A UN-Water Analytical Brief.* Hamilton: United Nations University – Institute for Water, Environment, and Health. http://public.eblib.com/choice/publicfullrecord.aspx?p=3286309.

———. 2014. "A Post-2015 Global Goal for Water: Synthesis of Key Findings and Recommendations from UN-Water." UN-Water. www.un.org/waterforlifedecade/pdf/27_01_2014_un-water_paper_on_a_post2015_global_goal_for_water.pdf.

———. 2016. "Integrated Monitoring Guide for SDG 6: Targets and Global Indicators." UN-Water. Available at: www.unwater.org/publications/sdg-6-targets-indicators/.

———. 2017. "Step-by-Step Monitoring Methodology for Indicator 6.5.2." UN-Water. Available at: www.unwater.org/publications/step-step-methodology-monitoring-transboundary-cooperation-6-5-2/

———. 2021. "Indicator 6.5.2 - Transboundary Water Cooperation." UN-Water SDG 6 Data Portal. 2021. www.sdg6data.org/indicator/6.5.2.

Wolf, Aaron T. 2007. "Shared Waters: Conflict and Cooperation." *Annual Review of Environment and Resources* 32 (1): 241–69. https://doi.org/10.1146/annurev.energy.32.041006.101434.

Wolf, Aaron T., and Joshua T. Newton. 2007. "Case Study of Transboundary Dispute Resolution: The Nile Waters Agreement." *Oregon State University, Corvallis.* http://transboundarywater.geo.orst.edu/research/case_studies/Documents/nile.pdf.

Wolf, Aaron T., Shira B. Yoffe, and Mark Giordano. 2003. "International Waters: Identifying Basins at Risk." *Water Policy* 5 (1): 29–60. https://doi.org/10.2166/wp.2003.0002.

Working Group on Integrated Water Resources Management. 2016. "Reporting under the Water Convention and Monitoring of Global SDG Indicator 6.5.2." WG.1/2016/INF5. Geneva: UNECE Convention on the Protection and Use of Transboundary Watercourses and International Lakes.

Wouters, Patricia. 2013. "International Law–Facilitating Transboundary Water Cooperation." *Published by the Global Water Partnership, TEC Background Papers*, no. 17. http://papers.ssrn.com/sol3/papers.cfm?abstract_id=2363809.

Zahid, Anwar and Syed Reaz Uddin Ahmed. 2006. "Groundwater Resources Development in Bangladesh: Contribution to Irrigation for Food Security and Constraints to Sustainability." In *Groundwater Research and Management: Integrating Science into Management Decisions. Proceedings of IWMI-ITP-NIH International Workshop on "Creating Synergy Between Groundwater Research and Management in South and Southeast Asia", Roorkee, India. 8–9 February 2005*, edited by Bharat R Sharma, Karen G Villholth, and Kapil D Sharma, 25–46. Groundwater Governance in Asia Series 1. Colombo, Sri Lanka: International Water Management Institute.

Zartman, I. William, ed. 2008. "Concept: Cooperation." *PIN Points* 30: 5–7.

Zeitoun, Mark, and Naho Mirumachi. 2008. "Transboundary Water Interaction I: Reconsidering Conflict and Cooperation." *International Environmental Agreements: Politics, Law and Economics* 8 (4): 297–316. https://doi.org/10.1007/s10784-008-9083-5.

Part II

Challenge 1

How is transboundary water cooperation defined?

4 A lack of consensus in defining transboundary water cooperation

Concise definitions have a place, as they are generalizable, and enable users to discuss concepts conceptually, but they prove difficult to apply and measure in practice. This challenge is inherent within the transboundary water cooperation literature, and I believe, a contributing factor to the number of definitions for transboundary water cooperation. Transboundary water cooperation can take many forms, modes, facets, and meanings, as noted by the data, which means that any definition developed will be broad or generic.

Transboundary water cooperation over internationally shared surface- and groundwaters is essential given the physical extent and global reliance on these waters. Increasing population growth, economic development, and climate change, as well as other factors, will only increase demand and competition for these shared resources, making cooperation an important process for increasing global access, meeting environmental needs, and encouraging positive interactions between states over water. Yet, despite the near-universal and, at times, common sense understanding that cooperation over transboundary waters is needed, we do not have a commonly accepted definition for what transboundary water cooperation is (Tarlock 2015; Saruchera and Lautze 2015). This is the first challenge addressed by this book. Through Chapters 4, 5, and 6, Part II answers the question: How is transboundary water cooperation defined?

This lack of a commonly understood definition poses problems when we attempt to measure progress toward cooperation. Depending on the definition or understanding of transboundary water cooperation, the method for measurement changes and the results vary widely. Chapter 3 presents this issue through the example of SDG Indicator 6.5.2 and the two alternative methods for measuring cooperation. Using this example, we can see the impact a relatively small change in the definition of cooperation and the measurement method can have in explaining the extent of cooperation in a country, river basin, or aquifer. Other researchers have noted this issue within the literature. For example, Woodhouse and Zeitoun (2008, 116) point out that the "term 'cooperation' has been equally loosely used. But 'cooperation' on a watercourse is far from being completely understood." Kistin (2007, 8) extends this further by discussing three considerations for understanding transboundary water cooperation that requires further research:

DOI: 10.4324/9781003126249-6

> The first consideration involves moving beyond the notion of cooperation as treaties to a more dynamic view of transboundary water cooperation as an ongoing and non-linear process in which state and non-state actors establish, challenge, modify and legitimize multi-layered governance structures. The second consideration reflects the relevance of scale for understanding the emergence, articulation, and effects of cooperation. The third involves dedicating more attention to the study of the dynamic effects and effectiveness of transboundary water cooperation.

Part II addresses this gap – beginning with this chapter – by reviewing the general cooperation and transboundary water cooperation literature and exploring several in-depth examples of existing methods used to measure transboundary water cooperation.

Cooperation in the broader literature

Cooperation is a term we regularly use in everyday life. We all inherently know the definition of cooperation, but we are likely to struggle when asked to define it. Merriam-Webster defines cooperation as: "the actions of someone who is being helpful by doing what is wanted or asked for: common effort; association of persons for common benefit" (Merriam-Webster 2021). Similarly, Oxford English Dictionary defines cooperation as: "the action of cooperating, i.e., of working together towards the same end, purpose, or effect; joint operation" (Oxford English Dictionary 2021). However, these definitions focus on the individual and are quite broad. In order to truly dive into what transboundary water cooperation is, we must first consider what cooperation means in other fields. The literature on cooperation is vast, and this section presents an overview of cooperation as it is defined by three fields of study: evolutionary and behavioral studies, philosophy, and international relations.

Cooperation is found at all levels – from the smallest biological units and organisms to individuals and states. Evolutionary explanations of cooperation define it as: "a behavior that provides a benefit to another individual (recipients), and the evolution of which has been dependent on its beneficial effect for the recipient" (West, Griffin, and Gardner 2007, 2). There are two general categories that explain the evolution of cooperation: direct fitness benefits and indirect fitness benefits. These draw from the theory that individuals can increase their fitness by directly passing on their own genes or through related individuals (Hamilton 1964). Individuals will cooperate to obtain direct fitness benefits, representing the personal fitness gained from an individual's behavior or reproduction. Indirect fitness benefits occur when a cooperative behavior benefits other individuals who often are related to the individual (West, Griffin, and Gardner 2007). Fitness benefits provide the foundation for the two general theories that explain cooperation. (1) *Kin selection* is the "process by which traits are favored because of their beneficial effects on the fitness of relatives" (West, Griffin, and Gardner 2007, 2). Cooperation is aimed at relatives with whom genes are shared. Expanding this theory, other scholars note that cooperative behavior due to kin selection draws on group altruism, where the individual does not benefit, but the group does (Zartman 2008; Axelrod and Hamilton 1981). Humans appear to be one species where group altruism occurs without high levels of genetic relatedness (Axelrod and Hamilton 1981). Altruism is defined as a behavior that is potentially costly to the actor, but beneficial to the recipient. (2) *Reciprocation theory* stems from symbiosis in nature; individuals will cooperate if there are expectations of reciprocity derived from past relationships of trust or from fear of future punishment or retaliation (Zartman 2008; Axelrod and Hamilton 1981). In either theory, individuals can benefit from mutual cooperation, or they can defect and exploit the cooperative efforts of other individuals. Having common interests as well as continuous contact can reduce this risk and encourage continued reciprocity (Axelrod and Hamilton 1981). These evolutionary theories provide one perspective on what cooperation is, which has similarities to philosophical perspectives.

Tuomela (2000) notes that there has been little comprehensive study of cooperation within the field of philosophy, despite the extensive work done in other fields. He cites that cooperation is based on kin and reciprocity; however, he extends these theories to explain the motivation background for cooperation. Friendship cooperation corresponds to *full-blown cooperation*, which "requires that participants are collectively acting towards the realization of a shared collective goal, collectively accepted by them as their goal" (Tuomela 2000, 369). Alternatively, *exchange cooperation* (or reciprocity cooperation) differs, as it refers to cooperation that is only based on shared private goals – coaction – rather than through friendship and cooperative attitude. Exchange cooperation occurs in more 'distant' group members who are acting together for mutual gains from a self-interest perspective (Tuomela 2000). Cooperation is joint action in either case; however, joint actions can be performed cooperatively and non-cooperatively. Cooperative joint action, according to Tuomela, occurs when participants are willing to complete not only their parts but also assist in additional activities needed for the joint action. Comparatively, non-cooperative joint action is when participants complete their agreed-upon part but will not assist in any additional activities, even those required for the agreed-upon joint action (Tuomela 1993). Therefore, he argues that for joint action to be cooperative, it depends on whether the attitude is cooperative or non-cooperative. A cooperative attitude is the willingness to participate and accept a reasonable share of the benefits from the joint action. Non-cooperative attitudes include free riding and reducing the actions required by an actor to complete the joint action while still attempting to maximize the share of the benefits (Tuomela 1993). One point Tuomela makes that I believe is particularly relevant for considering transboundary water cooperation through a philosophical lens is that a coerced participant can perform a fully cooperative joint action. The cooperative attitude needed for friendship cooperation or a fully cooperative joint action is not related to altruism or psychological attributes. A coerced person can willingly participate and accept additional responsibilities to complete a joint action from fear of punishment or other reasons (Tuomela 1993; 2000), which parallels arguments made regarding coercive cooperation and the influence of hegemons in transboundary basins (Zeitoun et al. 2019; Zeitoun and Allan 2008; Zeitoun, Mirumachi, and Warner 2020).

The international relations literature has its own perspectives on cooperation that focus on state-level interactions and the international system. This body of work is directly applicable to understanding transboundary water cooperation, given that they share a state-centric focus and that water is increasing in prominence in the international governance system. The international relations literature has generally developed a consensus on the definition of cooperation (Milner 1992). According to Milner, the most accepted definition is "when actors adjust their behavior to the actual or anticipated preferences of others through a process of policy coordination" (Milner 1992; Lindblom 1965, 227; Keohane 1984).

Within international relations literature, there are two primary schools of thought on cooperation: realism and liberalism. The realist theory is the most dominant. The main assumptions underlying realist views on cooperation are that people are inherently selfish and aggressive, meaning that state actors are driven by these motives and, therefore, conflict is inevitable (Clackson 2011). The primary goal of a state is to maximize its own power and security. Cooperation that occurs is short-term, and states are concerned with maximizing their share of the benefits, leading to a potential for defection and free riding (Zartman 2008). Furthermore, realists believe the international system is in a Hobbesian, anarchic state, as there is no authoritative government that can enforce the rules of behavior. Instead, it is often up to the hegemon to maintain stability and a cooperative order, also known as hegemonic stability theory (Herbert 1996; Zartman 2008).

Liberalism, on the other hand, developed following World War I and, according to which, the main obstacle to cooperation is that there are not enough international organizations to enforce the rules and norms (Clackson 2011). Liberals agree with realists in that the international system is in a state of anarchy but believe that war is not inevitable and there is potential for cooperation through democracy and international organizations (Zartman 2008; Keohane 1989). Keohane argues that cooperation can develop based on common interests between states; shared economic interests can create a demand for international institutions that states agree to follow (Keohane 1984). More recently, international relations literature has been expanding ideas on cooperation theory to include non-state actors, transnational norms and ideas, as well as the examination of effectiveness (O'Neill, Balsiger, and VanDeveer 2004). O'Neill, Balsiger, and VanDeveer (2004, 150) argue that cooperation's definition has been expanded from the "deliberate and coordinated adjustment of policies by states attempting to solve a mutual problem or achieve mutual gains" to the expanded definition where cooperation "comprises an iterated process, which continue beyond initial agreements and results in complex and enduring governance orders and potential social change." There are similarities between this definition and findings on transboundary water cooperation that will be discussed later in this chapter.

Table 4.1 Example of a range of definitions of cooperation from several fields

Cooperation definition	*Field and source*
Cooperation is defined as "the actions of someone who is being helpful by doing what is wanted or asked for: common effort; association of persons for the common benefit."	Dictionary (Merriam-Webster 2021)
Cooperation is "the action or process of working together to the same end; assistance, especially by complying readily with requests."	Dictionary (Lexico 2019)
Cooperation is "a behavior that provides a benefit to another individual (recipients) and the evolution of which has been dependent on its beneficial effect for the recipient."	Evolutionary Studies (West, Griffin, and Gardner 2007, 2)
Full-blown cooperation "requires that participants are collectively acting towards the realization of a shared collective goal, collectively accepted by them as their goal."	Philosophy (Tuomela 2000, 369)
"Cooperation is defined as a situation in which parties agree to work together at some cost to produce new gains for each of the participants that would be unavailable to them by unilateral action."	International Relations (Zartman 2008, 5)
Cooperation is a "deliberate and coordinated adjustment of policies by states attempting to solve a mutual problem or achieve mutual gains."	International Relations (O'Neill, Balsiger, and VanDeveer 2004, 150)
The expanded definition is where cooperation "comprises an iterated process, which continues beyond initial agreements and results in complex and enduring governance orders and potential social change."	International Relations (O'Neill, Balsiger, and VanDeveer 2004, 150)
Cooperation is "when actors adjust their behavior to the actual or anticipated preferences of others through a process of policy coordination."	International Relations (Keohane 1984; Lindblom 1965)

Transboundary water cooperation has not only been influenced by cooperation theories in other fields but also by applications of cooperation in customary and international law (see expanded discussion on international water law in Chapter 2). Overall, we can see a wide range of fields have addressed theories of cooperation and developed definitions. Table 4.1 presents a sample of definitions of cooperation, generally, in various fields.

Survey of existing measures of transboundary water cooperation[1]

In addition to developing definitions and theories on cooperation within and outside of the transboundary waters field, there has been a focus on measuring cooperation and tracking progress. Four studies stand out that have previously attempted to monitor transboundary water cooperation and map global efforts toward collaborative water governance, in addition to SDG Indicator 6.5.2 (discussed in Chapter 3). These studies further illustrate the challenges of defining and measuring transboundary water cooperation, which creates the potential for variation when different definitions are used. However, we can also use these to explore the common themes within prominent conceptualizations of cooperation.

Mapping the resilience of international river basins to future climate change-induced water variability

In 2010, the World Bank commissioned a report on the interactions of transboundary river basin management and climate change to better aid future design of cooperative measures that can adapt to climate variability and uncertainty: *Mapping the Resilience of International River Basins to Future Climate Change-Induced Water Variability*. The study's ultimate goal was to determine how resilience mechanisms, such as treaties and river basin organizations (RBOs), are globally distributed when compared to current and predicted hydrologic regime changes (De Stefano et al. 2010). To do this, the study's authors measured institutional capacity in comparison to climate risk. The existence of a treaty and/or an RBO has the potential to increase cooperation to a greater extent than in locations without any agreements or organizations. However, the mere presence of a treaty or RBO is not the only important factor. The mechanisms and design of the instruments and institutions are important, plus international water law can play a valuable role in reducing tension and adding resilience. The measurement for institutional vulnerability in this research essentially measured the level of transboundary cooperation within the river basin on a per basin-country basis. The criteria used to evaluate the institutional resilience of a basin-country area included the presence of a treaty, a water allocation mechanism, a mechanism for flow variability management, a conflict resolution mechanism, and the presence of an RBO. Each criterion was given equal weight; the fewer criteria met, the more vulnerable the basin-country area (De Stefano et al. 2010).

The methodology used in this study has several unique aspects. First, the scale of the World Bank study is at the level of a basin in a particular country. Second, the study looks at institutional vulnerability rather than transboundary cooperation. In a sense, the institutional vulnerability could be considered the lack of transboundary cooperation, and therefore the assumption put forward would be that a lower vulnerability score indicates a greater degree of cooperation. However, the substantive mechanisms identified as criteria were selected based on their ability to reinforce institutional resilience toward climate change. Lastly, the focus is on international river basins, thereby excluding transboundary aquifers, which are an important water resource globally. This study, however, still presents an alternative view of the institutional capacity that contributes to transboundary cooperation; it incorporates both substantive and procedural aspects, individual mechanisms, cooperative instruments, and institutions.

Water Cooperation Quotient

The second study is the Water Cooperation Quotient developed by the Strategic Foresight Group in 2013 as part of a study on water security in the Middle East (Strategic Foresight Group 2013). The authors have since refined the methodology and issued a report in 2015 titled *Water Cooperation Quotient* (Strategic Foresight Group 2015). The aim of this report is to present a means to measure the 'intensity and operational strength' of transboundary water cooperation. The authors argue that simply having a signed treaty that discusses allocation – one of the primary focuses of treaties on international waters (Hamner and Wolf 1997) – is not cooperation, but rather for cooperation to be operational, it needs to be *active* water cooperation. They define active water cooperation as "the commitment of countries to jointly manage their shared water resources" (Strategic Foresight Group 2015, 11). The Water Cooperation Quotient aims to measure active water cooperation by analyzing the water cooperation efforts of riparians, focusing on RBOs and formal agreements. The Quotient is composed of ten indicators that were given a rank score of 1 to 10, indicating the level of commitment to cooperation; the highest possible score is 55, which indicates the highest level of active water cooperation. The ten indicators used to calculate the Water Cooperation Quotient, in rank order from low to high, are existence of a formal agreement; existence of a river basin commission or organization; engagement at the ministerial level in cooperative meetings; collaboration in joint technical projects; joint environmental protection and quality control; joint monitoring of water flows; active collaboration in planned development, such as notification and consultation; commitment to cooperation at the highest political level; integration into regional economic cooperation; and evidence that cooperation mechanisms are functioning with active participation from the riparians (Strategic Foresight Group 2015).

From this, the authors put forward the bare minimum needed for a cooperative arrangement to be considered active water cooperation: the existence of a formal agreement; the existence of a river basin commission or organization; engagement at the ministerial level in cooperative meetings; and collaboration in joint technical projects. This study calculates the Quotient at a basin level, then averages the scores of the different basins within a country to yield a countrywide score. This can be disaggregated to examine the level of cooperation between a specific set of riparians. It includes a broad range of indicators and technical and economic factors, which are often not included in cooperation measures. The focus of the analysis, however, is strictly on the formal agreements and RBOs within a basin and does not include the potential for other forms of cooperation, such as through third-party or civil society actors.

Transboundary Waters Assessment Programme

The third assessment of global transboundary cooperation discussed is part of the UNEP and Global Environment Facility (GEF) *Transboundary Waters*

Assessment Programme (TWAP). The TWAP creates a baseline assessment of all the world's transboundary waters, including surface- and groundwater. The project's overarching aim is to establish a 'sustainable institutional framework' that can be used to develop a baseline assessment and a future assessment to measure changes in transboundary waters (UNEP-DHI 2016). Of relevance to our discussion is the Transboundary River Basins Assessment (TWAP-RB).[2] Like the overarching project, the river basins assessment also creates a baseline comparative assessment of the world's transboundary river basins by developing a methodology based on a range of issues, such as water stress, ecosystem threats, and the socioeconomic and governance capacity to cope with these issues. In addition, the governance capacity within the basins is considered at both national and international levels while aiming to identify the risk of interstate tension due to development or a lack of adequate institutional capacity. The governance thematic section of the TWAP-RB assessment includes three indicators: Legal Framework, Hydropolitical Tension, and Enabling Environment. The combination of these three indicators presents a framework and baseline for measuring not only transboundary cooperation but also the link between international cooperation and national policy.

Transboundary cooperation is captured by the Legal Framework and Hydropolitical Tension indicators. The Legal Framework indicator arises from the concept that legal agreements provide a framework for managing transboundary waters; therefore, it maps the spatial distribution of several fundamental international principles in water treaties (UNEP-DHI 2016). The fundamental principles are equitable and reasonable use; not to cause significant harm; environmental protection; cooperation and information exchange; notification, consultation, or negotiation; consultation and peaceful settlement of disputes; and whether a country has ratified either the UN Watercourses Convention or the UNECE Water Convention (UNEP-DHI 2016). The Hydropolitical Tension indicator considers the institutional resilience to development in the basin (De Stefano et al. 2017). The measurement of institutional resilience follows the calculation of institutional resilience in the aforementioned World Bank study, quantifying a score based on whether the basin-country area has a treaty, allocation mechanism, flow variability mechanism, conflict resolution mechanism, and/or a river basin organization. The level of institutional resilience is then compared to the planned, proposed, and under construction development of dams, reservoirs, and other major infrastructure projects (UNEP-DHI 2016; De Stefano et al. 2017). The Legal Framework indicator and the institutional resilience component of the Hydropolitical Tension indicator complement each other to effectively measure transboundary cooperation from a formal perspective. This perspective takes both substantive and procedural rules of customary international law into account, as well as formalized legal mechanisms specific to enhancing institutional capacity.

The third component of the governance thematic section of the TWAP-RB assessment measures the Enabling Environment; this indicator is similar to the proposed methodology for SDG Indicator 6.5.1.[3] This indicator is based on questionnaire data on policy, strategic planning, and legal frameworks; governance and institutional frameworks; and management instruments drawn from the 2012 IWRM Status Report (UNEP-DHI 2016; UNEP and UNDP 2012). The management of basins at the national scale has the potential to impact transboundary cooperation and vice versa. Therefore, the Enabling Environment indicator links the transboundary (or basin level) with the national level; this is important, as those countries struggling with implementing integrated water resources management at the national level may have less capacity to address issues and challenges at the transboundary level (UNEP-DHI 2016).

As with the other two studies discussed above, the TWAP-RB assessment only focuses on the transboundary cooperation within river basins. The TWAP project included a separate methodology for assessing the governance of transboundary aquifers; however, this differs from the methods used to assess river basins (UNESCO-IHP 2011).

The Blue Peace Index

The final individual assessment that presents a measure of transboundary water cooperation is the Blue Peace Index developed by the Economist Intelligence Unit. This index is a composite of five thematic areas, which enables it to "examine the extent to which countries and basins are managing their shared water resources in a sustainable, equitable and collaborative manner" (The Economist Intelligence Unit 2020, 4). While the Blue Peace Index does not explicitly measure transboundary water cooperation and has a broader focus on management, many of its component indicators address aspects of transboundary water cooperation.

The five domains of the Blue Peace Index are (The Economist Intelligence Unit 2020; 2019): (1) policy and legal frameworks, which assesses the national and basin-level policies, laws, and regulations; (2) institutional arrangements and participation – this domain evaluates the national- and basin-level institutional arrangements, the extent of stakeholder engagement, and data sharing; (3) water management instruments, which consider the extent to which national- and basin-level tools, such as pollution control, disaster management, monitoring and assessment, and allocation, have been developed and implemented in the basin or countries; (4) infrastructure and financing – this domain assesses the investment in water management at the national and international level, including indicators on public investment in water infrastructure, private sector investment, and investment in basin-level programs, such as for a joint institution or infrastructure; and (5) cooperation context – this final domain incorporates the broader socioeconomic, physical, geographic, and political context by including indicators on significant drivers and risk factors to peace. These include indicators on water stress, political stability, the potential for water-related conflict, as well as broader economic relations (The Economist Intelligence Unit 2019; 2020).

Thus far, the Blue Peace Index has assessed seven river basins: Amazon, Amu Darya, Mekong, Tigris-Euphrates, Sava, Senegal, and Syr Darya (The Economist Intelligence Unit 2020). It is limited in that it only considers groundwater when hydrologically connected to the river basins of evaluation (The Economist Intelligence Unit 2020); however, the methodology in principle could be applied to transboundary aquifers provided the data was available. Furthermore, as with the other measures, the index methodology has limited capacity to incorporate the intangible aspects of cooperation, such as trust and relationship building, and is primarily focused on state-level policies and programs, which could overlook cooperation and management occurring at subnational levels.

Other measures of transboundary water cooperation

Beyond the studies described above, Saruchera and Lautze surveyed past work measuring water cooperation and governance to develop a list of indicators. The purpose of this list was to advise on how transboundary water cooperation could be measured in the SDGs (2015). Their study identified six indicators for measuring cooperation: the existence of a transboundary agreement; reference to transboundary waters in national legislation; an inclusive basin plan; regular data exchange; standardized units and methods of measurement for water data; and financing available for transboundary institutions and projects (Saruchera and Lautze 2015). Three of these indicators are 'on paper' and the other three are based 'on practice' and therefore much harder to verify on a global scale, as they attempt to reflect the actual cooperation occurring. While it is too extensive to discuss all of the methods used to measure cooperation and its associated terms in as much detail, the following Table 4.2 lists reports, studies, and articles that measure cooperation, governance, management, or similar terms.

This table further illustrates the challenge addressed by Part II (see Chapter 2 for expanded discussion of the challenge) and captures a significant number of definitions and measurements related to transboundary water cooperation. When reading through this table, there are two overarching takeaways: (1) there is a high degree of variation between methods of conceptualizing and measuring transboundary water cooperation and (2) there is also significant overlap between transboundary water cooperation and similar terms, such as cooperative regimes, governance, and management priorities. The high degree of variability and the overlap with other terms could be due to limitations in data availability and the decision to develop a particular method and term to measure to fit the specific research or applied context. From a research perspective, having multiple definitions and tools for measurement can be valuable. It provides a means to triangulate and validate perspectives of transboundary water coordination, which is the method taken by this book in collecting and synthesizing existing research and methods for cooperation. Using these data from Table 4.2, we can see that the most common aspects of the definitions are the existence of a treaty, agreement, or formal institution and specific procedural and substantive mechanisms within these instruments. These same aspects

Table 4.2 List of terms and their descriptions that define or measure transboundary water cooperation or a similar term, including governance, management, or diplomacy

What is measured?	*Description*	*Source*
Water governance	Measurement of the Organization for Economic Cooperation and Development's (OECD) 12 Principles on Water Governance. These include roles and responsibilities, scale, cross-sectoral coordination, capacity, data and information, finance, implementation, trust, transparency, monitoring, and engagement.	(Akhmouch and Correia 2016; OECD 2018)
Cooperation as a dependent variable for water management	Existence of an agreement, treaty, or international institution.	(Bernauer and Siegfried 2008)
Degree of substantive international cooperation	The paper provides a summary of several studies that address the degree of substantive international cooperation qualitatively and by describing the compliance or the outcomes achieved over time.	(Bernauer and Siegfried 2008)
Performance of international cooperation	Policy performance metric based on time, optimum performance, actual performance, and counterfactual performance.	(Bernauer and Siegfried 2008)
International river basin cooperation	Existence of a treaty to facilitate cooperation, geographical configuration, and event data.	(Brochmann 2012)
Implementation of the Framework Agreement	Self-reporting mechanism with reports on implementation, programs, treaties and agreements, laws, institutions, and obstacles for cooperation.	("Decision on Adoption of the Methodology for Permanent Monitoring of Implementation of the Framework Agreement on the Sava River Basin" 2007)
Treaty cooperation	Variables include treaty existence and the number of treaties between riparians. Acknowledges other ways of measuring cooperation, such as the existence of joint organizations or commissions.	(Dinar et al. 2010)

(*continued*)

Table 4.2 Cont.

What is measured?	*Description*	*Source*
Blue Peace	"Refers to water cooperation across borders to foster stability and sustainable development. This can be in the form of shared institutions and legal frameworks which bring countries together in a commitment to resolve differences peacefully – and to use their shared water as a foundation for wider economic and diplomatic collaboration." Composite index based on multiple indicators in five thematic areas: policy and legal frameworks, institutions and participation, water management instruments, infrastructure and financing, and cooperation context.	(The Economist Intelligence Unit 2020, 15)
Principles of Effective Joint Bodies	A checklist for more sustained, mature cooperation. Includes establishment and function of joint bodies, operations, and financial and human resources.	(Honkonen and Lipponen 2018)
International Rivers Cooperation and Conflict Scale	The level of cooperation or conflict within an event or interaction between riparian states. Coding scale ranges from -6 to +6.	(Kalbhenn and Bernauer 2012)
Cooperation obligations in treaties	Regular information and data exchange, notification of planning and emergency measures, consultation, implementation, and joint mechanisms.	(Leb 2013)
Effectiveness of treaties in dispute avoidance and settlement	Evaluation of enforcement capability, treaty implementation, and dispute settlement mechanism.	(Ma et al. 2007)
Cooperation Intensity	Part of the Transboundary Waters Interaction Nexus (TWINs) Framework, ranges from confrontation, ad hoc, technical, risk averting, risk-taking. Identifies the existence of common goals, joint action, intention, and belief of collective action by another actor. Draws from idea of action-dependence (Tuomela 2000).	(Mirumachi 2007)
Good Transboundary Water Governance Matrix	A legal framework including substantive and procedural criteria.	(Orme et al. 2015)

Table 4.2 Cont.

What is measured?	*Description*	*Source*
Governance Regime, Performance, and Context	Specific characteristics and indicators are defined for three components of the framework: water governance regime, regime performance, and environmental and socioeconomic context.	(Pahl-Wostl et al. 2012)
Cooperative Regional Assessments	Include a transboundary analysis of benefits, a distributive analysis of the relative share of costs and benefits, and an institutional analysis of modes of cooperation.	(Sadoff and Grey 2005)
Transboundary water cooperation	Presence of a transboundary water agreement.	(Schweizerische Eidgenossenschaft 2013)
Legal framework for analyzing international water regimes	Five elements: scope, substantive rules, procedural rules, institutional mechanisms, and dispute settlement.	(Su 2014)
Transboundary water management priorities	Completion of an agreement and the creation of a river basin organization.	(Taylor 2013)
Good governance for transboundary freshwater and marine water bodies and groundwater systems	Eighteen criteria for reviewing the legal and institutional frameworks includes: legal basis, member states, scope, legal personality, functions, organizational structure, relationships, decision making, dispute resolution, data exchange, notification, finance, benefit-sharing, compliance, participation, and termination.	(UNDP-GEF 2011)
SDG Indicator 6.5.2 – operational arrangement for transboundary water cooperation	For an arrangement to be considered operational, it needs a joint body, regular meetings, joint plans, and regular data exchange.	(UN-Water 2016b)
Legal analytical framework of cooperative regimes	Scope, substantive rules, procedural rules, institutional mechanisms, and dispute settlement are addressed in agreements.	(Wouters and Chen 2013)
Conflict and Cooperative Events	The conflictive or cooperative intensity of an event – an interaction between riparian states.	(Yoffe 2001; Wolf, Yoffe, and Giordano 2003)
International river basin cooperation	Existence of a water-related treaty between riparian countries. Along with surface area, sources of freshwater, relative geographic power, economic and military capabilities.	(Zawahri and Mitchell 2011)

overlap with the governance terms, highlighting an emphasis on legal and institutional components of transboundary water cooperation in definitions.

This chapter corroborates insights from other researchers that cooperation is a complex term that is difficult to define. While this may seem obvious, I think it is important to note that "'cooperation,' it appears, is more complex than we generally tend to think" (Zeitoun and Mirumachi 2008, 303). This creates a challenge when we attempt to develop a comprehensive standardized definition of such a complex concept – one of the goals of Part II. Concise definitions have a place, as they are generalizable and enable users to discuss concepts conceptually, but they prove difficult to apply and measure in practice. This challenge is inherent within the transboundary water cooperation literature and, I believe, a contributing factor to the number of definitions for transboundary water cooperation. Transboundary water cooperation can take many forms, modes, facets, and meanings, as noted by the data, which means that any definition developed, will be broad or generic. The next chapter takes a deeper dive into definitions in both the literature and from interviews to synthesize the components – or 'key elements' – of transboundary water cooperation.

Notes

1 Text from this section is reproduced (with revisions and updates) from *Measuring transboundary water cooperation: options for Sustainable Development Goal Target 6.5* (Global Water Partnership, 2017) written by Melissa McCracken for the Global Water Partnership (GWP) for its Technical Committee Background Papers series.
2 *Transboundary River Basins: Status and Trends* (2016) and more information is available at: http://twap-rivers.org/. Data is available through the interactive data portal: http://twap-rivers.org/indicators/.
3 "Degree of integrated water resources management implementation," which is measured through a survey with questions surrounding four components: enabling environment, institutions, management instruments, and financing (UN-Water 2016a; 2016b).

References

Akhmouch, Aziza, and Francisco Nunes Correia. 2016. "The 12 OECD Principles on Water Governance –When Science Meets Policy." *Utilities Policy* 43 (December): 14–20. https://doi.org/10.1016/j.jup.2016.06.004.
Axelrod, Robert, and William D. Hamilton. 1981. "The Evolution of Cooperation." *Science* 211 (March).
Bernauer, Thomas, and Tobias Siegfried. 2008. "Compliance and Performance in International Water Agreements: The Case of the Naryn/Syr Darya Basin." *Global Governance* 14 (4): 479–501.
Brochmann, Marit. 2012. "Signing River Treaties—Does It Improve River Cooperation?" *International Interactions* 38 (2): 141–63. https://doi.org/10.1080/03050629.2012.657575.
Clackson, Alexander. 2011. "Conflict and Cooperation in International Relations." *E-International Relations* (blog). February 1, 2011. www.e-ir.info/2011/02/01/conflict-and-cooperation-in-international-relations/.

De Stefano, Lucia, James Duncan, Shlomi Dinar, Kerstin Stahl, Kenneth Strzepek, and Aaron Wolf. 2010. "Mapping the Resilience of International River Basins to Future Climate Change-Induced Water Variability." Water Sector Board Discussion Paper Series 15. Washington, D.C.: The World Bank.

De Stefano, Lucia, Jacob D. Petersen-Perlman, Eric A. Sproles, Jim Eynard, and Aaron T. Wolf. 2017. "Assessment of Transboundary River Basins for Potential Hydro-Political Tensions." *Global Environmental Change* 45 (July): 35–46. https://doi.org/10.1016/j.gloenvcha.2017.04.008.

"Decision on Adoption of the Methodology for Permanent Monitoring of Implementation of the Framework Agreement on the Sava River Basin." 2007. 5-07-16/3-3. Zagreb: First Meeting of the Parties to the Framework Agreement on the Sava River Basin.

Dinar, Ariel, Brian Blankespoor, Shlomi Dinar, and Pradeep Kurukulasuriya. 2010. "Does Precipitation and Runoff Variability Affect Treaty Cooperation between States Sharing International Bilateral Rivers?" *Ecological Economics* 69 (12): 2568–81. https://doi.org/10.1016/j.ecolecon.2010.07.036.

Hamilton, W.D. 1964. "The Genetical Evolution of Social Behaviour, I & II." *Journal of Theoretical Biology* 7: 1–52.

Hamner, Jesse, and Aaron T. Wolf. 1997. "Water: I. Patterns in International Water Resource Treaties: The Transboundary Freshwater Dispute Database." *Colorado Journal of International Environmental Law and Policy* 1997: 157–219.

Herbert, Anne L. 1996. "Cooperation in International Relations: A Comparison of Keohane, Haas and Franck." *Berkeley Journal of International Law*. https://doi.org/10.15779/z38135k.

Honkonen, Tuula, and Annukka Lipponen. 2018. "Finland's Cooperation in Managing Transboundary Waters and the UNECE Principles for Effective Joint Bodies: Value for Water Diplomacy?" *Journal of Hydrology* 567 (December): 320–31. https://doi.org/10.1016/j.jhydrol.2018.09.062.

Kalbhenn, Anna, and Thomas Bernauer. 2012. "International Water Cooperation and Conflict: A New Event Dataset." *SSRN Electronic Journal*. https://doi.org/10.2139/ssrn.2176609.

Keohane, Robert O. 1984. *After Hegemony: Cooperation and Discord in the World Political Economy*. Princeton: Princeton University Press. https://academic.oup.com/ia/article-lookup/doi/10.2307/2617490.

———. 1989. *International Institutions and State Power: Essays in International Relations Theory*. Boulder: Westview Press.

Kistin, Elizabeth J. 2007. "Trans-Boundary Cooperation in SADC: From Concept to Implementation." In *8th WaterNet/WARFSA/GWP-SA Symposium, Lusaka, Zambia*. Vol. 30.

Leb, Christina. 2013. "The UN Watercourses Convention: The Éminence Grise behind Cooperation on Transboundary Water Resources." *Water International* 38 (2): 146–55. https://doi.org/10.1080/02508060.2013.781474.

Lexico. 2019. "Cooperation." www.lexico.com/en/definition/cooperation.

Lindblom, Charles. 1965. *The Intelligence of Democracy*. New York: Free Press.

Ma, Jing, Keith W. Hipel, Mitali De, and Jun Cai. 2007. "Transboundary Water Policies: Assessment, Comparison and Enhancement." *Water Resources Management* 22 (8): 1069–87. https://doi.org/10.1007/s11269-007-9211-y.

Merriam-Webster. 2021. "Cooperation." In *Merriam-Webster*. Merriam-Webster. www.merriam-webster.com/dictionary/cooperation.

Milner, Helen. 1992. "International Theories of Cooperation Among Nations: Strengths and Weaknesses." *World Politics* 44 (3): 466–96. https://doi.org/10.2307/2010546.

Mirumachi, Naho. 2007. "Introducing Transboundary Waters Interaction Nexus (TWINS): Model of Interaction Dynamics in Transboundary Waters." In *Third International Workshop on Hydro-Hegemony*, 12–13. www.lmei.soas.ac.uk/water/publications/papers/file39696.pdf.

OECD. 2018. "OECD Water Governance Indicator Framework." OECD Water Governance Programme. www.oecd-ilibrary.org/environment/implementing-the-oecd-principles-on-water-governance/oecd-water-governance-indicator-framework_9789264292659-5-en.

O'Neill, Kate, Jörg Balsiger, and Stacy D. VanDeveer. 2004. "ACTORS, NORMS, AND IMPACT: Recent International Cooperation Theory and the Influence of the Agent-Structure Debate." *Annual Review of Political Science* 7 (1): 149–75. https://doi.org/10.1146/annurev.polisci.7.090803.161821.

Orme, Mallory, Zoë Cuthbert, Francesco Sindico, Julie Gibson, and Renice Bostic. 2015. "Good Transboundary Water Governance in the 2015 Sustainable Development Goals: A Legal Perspective." *Water International* 40 (7): 969–83. https://doi.org/10.1080/02508060.2015.1099083.

Oxford English Dictionary. 2021. "Co-Operation, n." In *Oxford English Dictionary Online*. Oxford University Press. www.oed.com/view/Entry/41037#eid8321561.

Pahl-Wostl, Claudia, Louis Lebel, Christian Knieper, and Elena Nikitina. 2012. "From Applying Panaceas to Mastering Complexity: Toward Adaptive Water Governance in River Basins." *Environmental Science & Policy* 23 (November): 24–34. https://doi.org/10.1016/j.envsci.2012.07.014.

Sadoff, Claudia W., and David Grey. 2005. "Cooperation on International Rivers." *Water International* 30 (4): 420–27. https://doi.org/10.1080/02508060508691886.

Saruchera, D., and J. Lautze. 2015. "Measuring Transboundary Water Cooperation: Learning from the Past to Inform the Sustainable Development Goals." International Water Management Institute (IWMI). www.iwmi.cgiar.org/publications/iwmi-working-papers/iwmi-working-paper-168/.

Schweizerische Eidgenossenschaft. 2013. "Swiss Position Paper on Water in the Post-2015 Agenda." Schweizerische Eidgenossenschaft. www.swisswaterpartnership.ch/wp-content/uploads/2014/07/2013.12.20_Swiss_Position_on_Water_in_Post2015_Agenda.pdf.

Strategic Foresight Group. 2013. *Water Cooperation for a Secure World: Focus on the Middle East*. Mumbai, India: Strategic Foresight Group.

———. 2015. *Water Cooperation Quotient*. Mumbai, India: Strategic Foresight Group. www.strategicforesight.com/publication_pdf/28799WCQ-web.pdf.

Su, Yu. 2014. "Contemporary Legal Analysis of China's Transboundary Water Regimes: International Law in Practice." *Water International* 39 (5): 705–24. https://doi.org/10.1080/02508060.2014.950856.

Tarlock, Dan. 2015. "Promoting Effective Water Management Cooperation Among Riparian Nations." *TEC Background Papers*, no. 21. www.gwp.org/Global/ToolBox/Publications/Background%20papers/GWP_TEC21_web.pdf.

Taylor, Paul. 2013. "National Stakeholder Consultations on Water: Supporting the Post-2015 Development Agenda." The Post-2015 Water Thematic Consultation. Stockholm: Global Water Partnership. https://sustainabledevelopment.un.org/content/documents/1815nationalstakeholder.pdf.

The Economist Intelligence Unit. 2019. "The Blue Peace Index 2019 - Methodology Note." London: The Economist. https://bluepeaceindex.eiu.com/pdf/Blue%20Peace%20Index%202019_methodology%20note.pdf.

———. 2020. "The Blue Peace Index 2020." London: The Economist. https://bluepeaceindex.eiu.com/pdf/EIU_Blue%20Peace%20Index_2020.pdf.

Tuomela, Raimo. 1993. "What Is Cooperation?" *Erkenntnis* 38 (1): 87–101. https://doi.org/10.1007/BF01129023.

———. 2000. *Cooperation*. Philosophical Studies Series 82. Dordrecht, Netherlands: Kluwer Academic Publishers.

UNDP-GEF. 2011. "International Waters: Review of Legal and Institutional Frameworks." Washington, D.C.: UNDP-GEF International Waters Project.

UNEP-DHI. 2016. "Transboundary River Basins: Status and Trends." Volume 3: River Basins. Transboundary Waters Assessment Programme. Nairobi, Kenya: United Nations Environment Programme.

UNESCO-IHP. 2011. "Methodology for the GEF Transboundary Waters Assessment Programme: Volume 2 Methodology for the Assessment of Transboundary Aquifers." Nairobi, Kenya: UNEP. www.geftwap.org/publications/methodologies-for-the-gef-transboundary-assessment-programme-1/volume-2.

United Nations Environment Programme and United Nations Development Programme (UNDP). 2012. *Status Report on the Application of Integrated Approaches to Water Resources Management*. Nairobi, Kenya: United Nations Environment Programme.

UN-Water. 2016a. "Draft Step-by-Step Monitoring Methodology for Indicator 6.5.1." UN-Water. www.unwater.org/fileadmin/user_upload/unwater_new/docs/DRAFT%20Step-by-step%20methodology%20for%206%205%201%20incl%20questionnaire_v2016-04-20.pdf.

———. 2016b. "Integrated Monitoring Guide for SDG 6: Targets and Global Indicators." UN-Water.

West, Stuart A., Ashleigh S. Griffin, and Andy Gardner. 2007. "Evolutionary Explanations for Cooperation." *Current Biology* 17 (16): R661–72. https://doi.org/10.1016/j.cub.2007.06.004.

Wolf, Aaron T., Shira B. Yoffe, and Mark Giordano. 2003. "International Waters: Identifying Basins at Risk." *Water Policy* 5 (1): 29–60. https://doi.org/10.2166/wp.2003.0002.

Woodhouse, Melvin, and Mark Zeitoun. 2008. "Hydro-Hegemony and International Water Law: Grappling with the Gaps of Power and Law." *Water Policy* 10 (S2): 103–19. https://doi.org/10.2166/wp.2008.209.

Wouters, Patricia, and Huiping Chen. 2013. "China's 'Soft-Path' to Transboundary Water Cooperation Examined in the Light of two UN Global Water Conventions - Exploring The 'Chinese Way.'" *The Journal of Water Law* 22: 19.

Yoffe, Shira. 2001. "Basins at Risk: Conflict and Cooperation over International Freshwater Resources." Corvallis, Oregon: Oregon State University. http://ir.library.oregonstate.edu/jspui/handle/1957/9693.

Zartman, I. William, ed. 2008. "Concept: Cooperation." *PIN Points* 30: 5–7.

Zawahri, Neda A., and Sara McLaughlin Mitchell. 2011. "Fragmented Governance of International Rivers: Negotiating Bilateral versus Multilateral Treaties." *International Studies Quarterly* 55 (3): 835–58. https://doi.org/10.1111/j.1468-2478.2011.00673.x.

Zeitoun, Mark, and John Anthony Allan. 2008. "Applying Hegemony and Power Theory to Transboundary Water Analysis." *Water Policy* 10 (S2): 3–12. https://doi.org/10.2166/wp.2008.203.

Zeitoun, Mark, and Naho Mirumachi. 2008. "Transboundary Water Interaction I: Reconsidering Conflict and Cooperation." *International Environmental Agreements: Politics, Law and Economics* 8 (4): 297–316. https://doi.org/10.1007/s10784-008-9083-5.

Zeitoun, Mark, Naho Mirumachi, and Jeroen Warner. 2020. *Understanding Water Conflicts: Analysis for Transformation*. New York: Oxford University Press.

Zeitoun, Mark, Naho Mirumachi, Jeroen Warner, Matthew Kirkegaard, and Ana Cascão. 2019. "Analysis for Water Conflict Transformation." *Water International*, May, 1–20. https://doi.org/10.1080/02508060.2019.1607479.

5 Similar terms

Teasing out definitions of transboundary water cooperation

Now that we have some conceptualization of the means and types of methods that measure transboundary water cooperation,[1] this chapter will shift to analyzing existing definitions and exploring the complexity present in the concept. Part II aims to answer the question: How is transboundary water cooperation defined? This goal can be understood in a couple of different ways. First is the approach taken in Chapter 4, which reviews the literature on general cooperation and transboundary water cooperation to see how cooperation has been defined in the literature. Second is the approach taken by this chapter to conduct an explicit analysis of a sample of definitions to generate a clear, consistent definition of transboundary water cooperation. To create this definition, we must understand the range of definitions and identify the core components most consistently identified as comprising cooperation. These core components – or 'key elements' – are identified using a qualitative content analysis of expert[2] interviews, peer-reviewed articles, and grey literature conducted and sampled by the author. These data were categorized into both predetermined and organically determined categories and themes – referred to as *codes*. The codes identified include implicit and explicit definitions, key elements of cooperation, methods to measure cooperation, and related terms such as management or governance.

Within the literature sample (n = 273) and interviews (n = 54), I coded 51 direct, explicit definitions of cooperation and 54 direct answers to the questions 'How might you define transboundary water cooperation? What are the key elements?' Definitions, when clearly stated, ranged from general to very detailed. While most interviewees defined the term, many took pause with the question, commented on its difficulty, or noted the challenges of defining it. Definitions varied in both the literature and interviews; the variations were often based on a specific perspective, bias, or context. Therefore, the definitions were coded for 'key elements' of cooperation. A key element of cooperation is a word or phrase representing a standalone concept or component of cooperation. A complex phrase or sentence may have more than one key element embedded and, therefore, could be coded more than once for different key elements. In total, I identified 50 key elements within the definitions of cooperation in both the literature and interviews. Key elements include concepts such

DOI: 10.4324/9781003126249-7

as basin planning, infrastructure, treaties or agreements, river basin organizations (RBOs), and adaptation. For a complete list of the key elements identified, see Appendix A. While a whole chapter could be devoted to each of these key elements and what they mean for cooperation, this is outside the scope of the book. The goal here is to define transboundary water cooperation clearly and consistently; therefore, this chapter presents only the main themes identified during the analysis. Overall, cooperation is more often left undefined in the literature, which contributes to the issues this book is attempting to address, and there is confusion between commonly used terms in the field – such as cooperation, management, governance, and diplomacy.

Explicit and implicit: How is cooperation discussed?

Since transboundary water cooperation is difficult to define, this may contribute to the lack of definitions within the literature. Within the data, definitions of cooperation can be separated into two categories – explicit and implicit discussions of cooperation. I define explicit discussions as the article text or the interview transcript that directly discusses a framework for cooperation, examples of cooperation, or directly defines the term. Implicit discussions are the text that addresses cooperation indirectly, such as text that discusses a case study and mentions elements of cooperation that are relevant or important to that case study. For example, the following text is coded as implicit (Honkonen and Lipponen 2018, 321):

> This should hold with the understanding that cooperation practices have developed in particular local contexts. The key features of cooperation arrangements that have brought about sustained cooperation, dialogue and the resolution of various issues, and which have also been adapted to deal with emerging challenges, merit attention.

Cooperation is not defined in this quote, but key elements, such as dialogue, context, and adaption, are mentioned; cooperation is discussed but not defined. Out of the 273 articles and grey literature reviewed, 74 discuss cooperation implicitly and 29 define cooperation explicitly. Some articles are coded for both implicit and explicit discussions.

Despite the significant body of literature on transboundary water cooperation, it is surprising to see the limited number that discusses a possible definition – implicitly or explicitly – of cooperation in this sample. Perhaps unsurprisingly, there are significantly more implicit than explicit discussions of cooperation. This same conclusion cannot be drawn from the interviews, as the interviewees were directly asked to define transboundary water cooperation; therefore, all discussed cooperation explicitly. Within the 29 articles with explicit discussions of cooperation, definitions of cooperation are further broken down[3] by how they are discussed by separating them into five categories: general cooperation definitions, direct definitions, examples, frameworks, and antithesis definitions.

From the literature sample, there are 11 explicit definitions of cooperation in a general sense, exclusively drawing from international relations literature and there are six definitions that provide examples. Six articles use a framework for cooperation as the means to define it, and there are two definitions of cooperation that state what cooperation is not but do not cite what cooperation is. Lastly, only 23 clear, direct, and explicit definitions of transboundary water cooperation were found in the literature. Of the 54 interviews, only 21 individuals stated a clear, direct, and explicit definition of cooperation. This is despite responding to a direct question asking them to define cooperation. I believe there are two main reasons for this discrepancy: (1) The question was a two-part question asking for the definition and the key elements. Every person was able to list and discuss key elements; however, this could have caused an omitted or overlooked response to the first part of the question. (2) There is not a single accepted definition of cooperation in the field, meaning that defining cooperation is a challenging task that has the potential to vary drastically depending on the scale, context, and place, as well as the background of the individual. Table 5.1 provides a sample of the explicit definitions to demonstrate the differences in how definitions of cooperation are constructed.

Table 5.1 Sample of explicit definitions of cooperation, separated by the five different methods for defining cooperation

Explicit definitions of cooperation	
General definitions of cooperation	
"Keohane (1984) describes cooperation as a process of policy coordination through which actors adjust their behavior to the preferences of others."	(Daoudy and Kistin 2008)
"Cooperation has been defined as a 'process by which States take coordination to a level at which they work together to achieve a common purpose that produces additional mutual benefits that otherwise would be unavailable with unilateral action.' (Zartman 2008)"	(de Chaisemartin 2018)
Examples of cooperation	
"Cooperation with Kazakhstan takes the form of setting up institutions, such as river commissions, and agreements to manage their shared water resources."	(Ho 2017)
"This Commission may well represent the best current example of functional cooperation – joint management and operation of transboundary resources between two highly dissimilar countries (Mumme 1993)."	(Shmueli 1999)
Frameworks of cooperation	
"Conceived thus, for an international institution or interaction to be considered an instance of 'cooperation' it must meet three conditions . . . policy coordination, mutual adjustments, and joint gains."	(Selby 2013)

(*continued*)

Table 5.1 Cont.

Explicit definitions of cooperation	
"The conceptual building blocks for a framework for understanding how cooperation over shared water resources 'works.' These building blocks are a) assessment of the river basin; b) contextual factors related to action situation; c) the institutions structuring action; d) the actors; e) the interface between structure-agency interface: the (action) situations in which different actors interact; and finally f) the different outputs and outcomes."	(Huntjens et al. 2016)
Antithesis definitions	
"Merely signing treaties for allocation of water resources between riparian countries is not cooperation. Even signing treaties which go beyond allocation and provide for exchanges and joint ventures are also not cooperation. For cooperation to be meaningful, it must be active in an operational way."	(Strategic Foresight Group 2015)
Direct explicit definitions – literature	
"Water cooperation relies on the fact that many watercourses are transboundary, and it ensures that the benefits that can be derived by managing such water resources cooperatively rather than unilaterally are generated for riparian communities and states."	(Schmeier 2018)
"Cooperation can be reflected in signing new treaties in cases where they do not exist; in more treaties to amend the initial set of agreements; or in new treaties introducing more issues (such as water quantity, hydropower, and flood control) in the cooperative framework."	(Dinar et al. 2010)
Direct explicit definitions – interviews	
"I would say that transboundary water cooperation happens whenever positive interactions take place between two or more actors involved in the use of resources that cross boundaries, whether they are international (e.g., between two countries), national (e.g., between two federal states) or local (e.g., between two villages or even sectors of activities)."	(Salamé 2018)
"I would define transboundary water cooperation or collaboration as the equal participation in decision making for shared waters or shared resources . . . on both sides."	(Schroeck 2018)

Common key elements of cooperation

In addition to how definitions are constructed and whether discussions are implicit or explicit, we can identify trends, similarities, and differences. These implicit and explicit discussions of cooperation were coded for key elements of cooperation. The following Table 5.2 lists the top five most commonly identified key elements in the literature and interviews, separated by whether the discussion of cooperation was implicit or explicit.

Table 5.2 Top five most commonly referenced key elements of cooperation broken down between the implicit and explicit definitions, as well as between the literature and interview data

Cooperation implicit – literature only	*Cooperation explicit – literature only*	*Question defining cooperation – interviews only*	*Cooperation explicit – literature and interviews combined*
Treaty or agreement	Treaty or agreement	Mutual benefit	Treaty or agreement
Institutional capacity	Institutional capacity	Data and information	Mutual benefit
Mutual benefit	Mutual benefit	Basin planning	Institutional capacity
Equitable	RBO	Institutional capacity	Data and information
Contextual	Stakeholder participation	Shared understandings	Basin planning

Overall, the three most commonly coded key elements are *treaty or agreement*, *mutual benefit*, and *institutional capacity*. Definitions referencing a treaty or agreement, or stating that a treaty or agreement is needed to formalize a relationship over shared waters, were coded as *treaty or agreement*. Treaties have historically been used as a measure of cooperation in the field of transboundary water and other fields. Data coded as *mutual benefits* includes text that directly references mutual benefits, mutual satisfaction of outcomes, or shared or equitable benefits. These terms have a substantial history in the field, which supports their prominence in the data. The idea of mutual benefits is prevalent as a core idea of cooperation in other fields outside of international waters, such as evolutionary studies and biology, where a mutual benefit is defined as "a benefit to both the actor and the recipient" (West, Griffin, and Gardner 2007, 2). This definition of mutual benefit mirrors definitions of mutual benefit or benefit sharing in transboundary waters literature: "as any action designed to change the allocation of costs and benefits associated with cooperation" (Sadoff and Grey 2005, 422).

Institutional capacity includes text that specifically mentions the term 'institutional capacity,' or referenced the combination of agreements, RBOs, or discussed institutions more generally. This differs slightly from the common definition of institutional capacity in the literature as the treaties, agreements, RBOs, other joint management institutions, and even positive international relations between nation-states (Wolf 2007; Wolf, Yoffe, and Giordano 2003). I separate formal institutional capacity – treaties, institutions, etc. – from the positive international relations. This was coded separately when mentioned in the data. The importance of institutional capacity and its prominence could stem from two sources. First is the concept of liberalism (aka institutionalism) from international relations literature, which states that international institutions and organizations promote cooperation and prevent conflict within the international system (Keohane 1989; 1984). This is supported in the field of international waters by the Basins at Risk project, which found that "the likelihood and intensity of dispute rises as the rate of change within a basin exceeds the institutional capacity to absorb that change" (Wolf, Yoffe, and Giordano 2003, 51). Therefore, cooperation, when considered as the opposite of conflict, is more likely when there is adequate institutional capacity.

Aside from the overlap in the top key elements between the literature and interviews as well as implicit and explicit discussions (see Table 5.2), we can see differences in the other top key elements coded in the definitions of cooperation. Implicit discussions in the literature also tended to include references to *equitable*. Predominately, these data refer to the customary and codified law of 'equitable and reasonable use,' but also address more general statements of equity or equitable allocation of water as well as benefits, and a few cases of equitable relationships in terms of power. In addition, implicit definitions commonly were coded as *contextual*. Data coded as *contextual* discusses the impact and/or importance of context on cooperation, including geography, politics, history, culture, and timing. Explicit definitions in the literature include *RBOs* and *stakeholder participation* instead. Data coded as *RBOs* include requirements for an RBO, commission, or some variation of a formal body that manages, governs, oversees, or monitors the basin. This aligns with the definition of RBOs developed by Schmeier, Gerlak, and Blumstein (2016, 4) as "institutionalized forms of cooperation that are based on binding international agreements covering the geographically defined area of an international river or lake basins characterized by principles, norms, rules and governance mechanisms." *Stakeholder participation* is defined broadly and includes all relevant riparians, non-state actors, and public participation in the process of cooperation.

Lastly, we see the greatest differences in the results from the academic and practitioner interviews. *Mutual benefit* and *institutional capacity* are still included as top key elements of cooperation; however, the other three differ from the literature: *data and information*, *basin planning*, and *shared understandings*. Data coded at *data and information* address joint collection of data, data and information exchange, and the need for quality data and information in the cooperative process. There is not any restriction of the scope or regularity of data and information exchange and collection to be included in this category. *Basin planning* includes references to the development of joint objectives, basin management plans, collaborative planning, and similar variations that were aimed at more technical or management-oriented collaboration. *Shared understandings* as a key element include identifying common interests, shared goals, and the more difficult to define concept of understanding and accepting another party's interests, and working to help achieve not only your goals but also theirs. While there is no definitive way to determine the cause or reasoning behind these differences in the definitions, I hypothesize that the key elements most commonly cited in the explicitly defined literature are more normative and more thoroughly understood. The idea of equitable, the role of context, and the idea of shared understandings are more difficult to generalize with respect to cooperation and, therefore, it is more difficult to measure and support their inclusion into a definition of cooperation. However, this does not mean that these concepts are not important for understanding and developing cooperation over shared waters. These are not the only key elements that are more difficult to measure or identify in a cooperative process that are mentioned within definitions of cooperation. Therefore, when developing a definition for transboundary water cooperation, it must be inclusive of these more intangible elements of cooperation.

Blurred lines: Differentiating between cooperation, management, governance, and diplomacy

Part II aims to define transboundary water cooperation; however, part of the difficulty in doing so lies with the blurred lines dividing transboundary water cooperation and other terms in the field, namely, water diplomacy, water governance, and water management. Previous authors have noted that there are multiple definitions or a lack of agreement over definitions of these terms in the literature. For example, Schmeier (2018) published an article about defining water diplomacy that addresses the differences between cooperation and management. Pohl et al. (2014) noted how water diplomacy has multiple definitions by various authors, and Molnar et al. (2017) conducted a comprehensive review of both water diplomacy and water cooperation definitions. It is not just diplomacy and cooperation that are often used interchangeably. An article by Jägerskog on transboundary water management devotes several sections to discussing the importance of cooperation and how to improve it (Jägerskog, n.d.). These are not isolated examples; these terms are related and interlinked, making differentiating between them difficult. This section identifies some of the similarities and differences between the definitions of the terms that became apparent during the data analysis. One interviewee notes that cooperation is difficult to define, and one place to start would be to "differentiate it from the other concepts that are close to it. For example, a lot of people are talking about transboundary water management or diplomacy. And so, I would probably try to identify the elements of transboundary cooperation by comparison." The intent of this section is not to establish a judgment on a specific definition of diplomacy, governance, or management, as that is outside the scope of this book. However, I do aim to help identify the links and differences between these terms and cooperation, so that we can better understand the definition of transboundary water cooperation.

Water diplomacy

There is no singular definition of water diplomacy; it is defined differently by various actors in different contexts – much like cooperation (Huntjens et al. 2016; Pohl et al. 2014; Al-Saidi and Hefny 2018; Klimes et al. 2019).

Cooperation and diplomacy are often used interchangeably, and by comparing definitions of diplomacy with the key elements of cooperation as identified in this book, we can identify places where they overlap. Water diplomacy has been defined as:

- "Water diplomacy includes all measures by state and non-state actors that can be undertaken to prevent or peacefully resolve (emerging) conflicts and facilitate cooperation related to water availability, allocation or use between and within states and public and private stakeholders." (Huntjens et al. 2016)
- "Water diplomacy can be defined as the use of diplomatic instruments to existing or emerging disagreements and conflicts over shared water resources with the aim to solve or mitigate those for the sake of cooperation, regional stability, and peace." (Schmeier 2018)
- "A theory and practice of implementing adaptive water management for complex water issues. The water diplomacy approach diagnoses water problems, identifies intervention points, and proposes sustainable solutions that are sensitive to diverse viewpoints and values, ambiguity and uncertainty as well as changing and competing needs." ("Water Diplomacy@ Tufts" 2019)
- "Water diplomacy is a dynamic and evolving process and can support important efforts to achieve *peaceful, inclusive and sustainable water cooperation* between communities, regions, and countries. It can be tailored and responsive to the uniqueness of each process and geopolitical context." (SIWI 2016)
- "Concepts of (water) diplomacy seek to help national states achieve mutually beneficial agreements through non-violent measures." (Al-Saidi and Hefny 2018)
- "Since one of the goals of water diplomacy is to strengthen the viewpoint of mutually shared benefits in bilateral and basin-wide contexts (Keskinen et al. 2013), our results confirm that water diplomacy cannot and does not focus only on water." (Yasuda et al. 2018)
- "Water diplomacy is an approach that enables a variety of stakeholders to assess ways to contribute to finding solutions for joint management of shared freshwater resources. It is a dynamic process that seeks to develop reasonable, sustainable and peaceful solutions water management while promoting or informing cooperation and collaboration among riparian stakeholders." (Klimes et al. 2019)

Looking at the definitions listed above, there is overlap with how cooperation is defined, and we can see similarities with the top key elements of cooperation identified in Table 5.2. Both terms are context-dependent and adapt to the geopolitical, cultural, and environmental context. Furthermore, the diplomacy definitions above cite the importance of various actors, including state- and non-state actors, communities, regions, and public and private stakeholders.

Interviews highlighted the importance of stakeholder participation in the cooperative process. Lastly, the aims of cooperation are commonly identified as the reduction or prevention of conflict and the achievement of mutual benefits, which align with many of the goals of water diplomacy. There is a fine line between diplomacy and cooperation, and the terms are intertwined, as diplomacy is used to facilitate or induce cooperation, and cooperation can lead to improved water diplomacy. Diplomacy is seen as high politics, whereas cooperation occurs at a lower or more technical level (Keskinen et al. 2013; Molnar et al. 2017). In addition, diplomacy is seen as having goals that are broader in scope than those of cooperation, such as regional peace, stability, or cooperation (Molnar et al. 2017; Schmeier 2018; 2016), whereas cooperation's goals are targeted to outcomes directly related to water.

Global water governance

Water governance, as you can surmise, also needs to be "carefully defined, as it may not be readily understood" (Rogers and Hall 2003); there is a lack of deep understanding of the dynamics of water governance (Turton and Earle 2005; Phillips et al. 2006). Water governance has a basis in governance literature. Governance has evolved, much as water governance has. Early understandings of the term equate its purpose with management (Villholth et al. 2018). Central to the definition of governance is the political power of the state to manage (World Bank 1989). Even though this has changed, the World Bank still asserts that governance is the process that gives rulers the authority to make, enforce, and modify rules (World Bank 2015 in Akhmouch and Correia 2016). The definition of governance has broadened to be more about the process, rather than the structure of management (i.e., government) and has moved away from a focus on state-centric power to a wider variety of actors that play a role in governing a narrower scope of governance (i.e., water) (Villholth et al. 2018). Water governance has embodied these changes in governance; common definitions of water governance include:

- "Water governance refers to the range of political, social, economic and administrative systems that are in place to develop and manage water resources, and the delivery of water services, at different levels of society." (Rogers and Hall 2003, 7)
- "Water governance formally refers to the range of political, institutional and administrative rules, practices, and processes (formal and informal) through which decisions are taken and implemented, stakeholders can articulate their interests and have their concerns considered, and decision-makers are held accountable for water management." (OECD 2015)

One key difference between these definitions and cooperation is that the focus on water governance is not strictly limited to international transboundary basins. While the focus on state power has shifted, much of the emphasis is still on the national governance of waters at the state, regional, and/or local level, incorporating a wide range of stakeholders. A subdivision of water governance (and the greater governance literature) has developed to focus specifically on global and inter-state management. Global governance stems from the idea that state power and authority are shifting toward international institutions (Finger, Tamiotti, and Allouche 2006) – trending with liberalism in international relations literature. Global water governance, therefore, considers the globally common norms and understandings of water management (Gupta and Pahl-Wostl 2013). Wiegleb and Bruns (2018) define global water governance as "the formal and informal institutions, activities, and processes between different actors through which collective interests on water are articulated, differences are mediated, and joint actions are established." The GEF has also promoted the concept of 'global international waters governance' through their Transboundary Diagnostic Analysis-Strategic Action Programme approach, which has shifted governance from more normative conceptualizations of treaties and agreements to a more 'action-oriented' process (Chen, Pernetta, and Duda 2013).

If the key elements of cooperation have a broader inclusion of stakeholders, institutional capacity, and shared understandings, which parallel definitions of global water governance, how then does global water governance differ from transboundary water cooperation? Daoudy and Kistin (2008) state that cooperation is the 'negotiated system for governance,' perhaps recognizing that cooperation plays a role in developing and improving water governance (Chikozho 2014). However, Yeophantong (2017) points out that cooperation is often seen as a means to improve water governance, but given the nature of the international water governance system, cooperation could occur in less than desirable governance systems, such as hegemonic relationships between states that could produce cooperative outcomes that are less than equitable. So perhaps the most notable difference between cooperation and global water governance is its scope. Global water governance, which focuses on water issues – unlike diplomacy, which has broader aims – encompasses a broader system for decision-making, such as domestic policy as well as cultural norms and societal values that contribute to how water is managed. Cooperation, on the other hand, focuses on the factors that contribute to the interactions between actors within a governance process as well as the outcomes of that process.

Water management

Historically, 'conventional' water management has been based on technical solutions that are rooted in the 'command and control' paradigm. This assumes that the behavior of the surface- or groundwater system is predictable, and any uncertainties can be managed or engineered out of the system through technical solutions and large infrastructure (Pahl-Wostl, Jeffrey, and Sendzimir 2011; Pahl-Wostl, Kabat, and Möltgen 2008). However, with climate change, rapid population growth, and development, the stationarity this conventional water management was built on is no longer valid (Milly et al. 2008). Definitions and practical applications of water management must change. Water management is a broad, interdisciplinary field and must encompass a wide range of topics, including drinking water, sanitation, and flood protection, among others. This wide range makes defining the term ambiguous (Global Water Partnership 2000; Grigg 2015). Water management has been defined as the "activities of analyzing and monitoring, developing and implementing measures to keep the state of water resources within desirable bounds" (Pahl-Wostl et al. 2012, 25). Alternatively, river basin management has expanded in scope to be defined as "the management of water systems as part of the broader natural environment and in relation to their socio-economic environment" (Mostert et al. 2000; Svendsen 2005). This broadening of scope is part of the paradigm shift in water management, which has seen the development of new conceptualizations. Adaptive water management is defined as "a systematic process for improving management strategies and by taking into account changes in external factors in a proactive manner" (Pahl-Wostl, Jeffrey, and Sendzimir 2011) and is a method for adapting to water-related uncertainties in the future since we can no longer rely on the stationarity of water systems. The most prominent development in water management is the idea of integrated water resources management (IWRM). While the ideas behind IWRM have been around for nearly 100 years (Giordano and Shah 2014), IWRM has come to the fore in the last 30 years with the development of the Dublin principles, which are at the core of IWRM (International Conference on Water and the Environment 1992; Pahl-Wostl, Jeffrey, and Sendzimir 2011; Global Water Partnership 2000). IWRM is defined by the Global Water Partnership as "a process which promotes the coordinated development and management of water, land, and related resources, in order to maximize the resultant economic and social welfare in an equitable manner without compromising the sustainability of vital ecosystems" (Global Water Partnership 2000, 22). These definitions of water management differ from cooperation in their inter-sectoral scope and focus on technical solutions. Like governance, water management, including IWRM, does not focus specifically on international transboundary basins. This poses a challenge for IWRM, as cooperation is needed for there to be integrated management (Wolf 2010); states can be reluctant to integrate at the international level, as that can mean a loss of sovereignty (Phillips et al. 2006). Therefore, a new subfield of water management specific to international shared waters has developed, which can be compared with cooperation. Transboundary water management or international river basin management can be defined as:

- "Transboundary water management can be defined as the application of technical tools to specific water-related questions and challenges to improve water-specific outcomes in watercourses that are transboundary by nature." (Schmeier 2018)
- "In this context, we understand 'river basin management' in a broad sense, including not only river basin management plans themselves, but all actions and measures taken that aim to avoid or mitigate conflict over shared water

resources, while increasing the benefits of cooperation and sharing them across the basin." (Schmeier and Vogel 2018)

Other authors have further developed the best practices and principles of transboundary water management. For example, Van der Zaag and Savenije (2000) identify integrated supply and demand management, public participation, and enhanced regional and economic integration as best practices. This differs from the international legal principles that Rahaman cites as central to transboundary water management, which includes limited territory sovereignty, equitable and reasonable use, and no significant harm, among the other central tenets of international water law and the 1997 UN Watercourse and 1992 UNECE Water Conventions (Rahaman 2009; United Nations Economic Commission for Europe 2013; United Nations 1997).

Between the conventional definitions of water management, IWRM, and transboundary water management, there is a breadth of definitions and understanding of water management. Some definitions, such as the definition of river basin management by Schmeier and Vogel (2018), have more in common with the key elements of cooperation than others. From this analysis, one of the main differences between water management and cooperation is that management is more technical and includes the day-to-day activities that are needed for management and cooperation over shared water resources, such as monitoring, implementation, or joint infrastructure development and operation. However, these activities play an important role in building trust, working relationships, and shared understanding in order to foster cooperation between actors. One interviewee stated that management is the outcomes, while cooperation is the decision-making process that enables management to determine the appropriate outcomes and implement them. Notably, this differed with the majority of the other interviewees, who generally equated outcomes with effective cooperation, which is discussed in Part III.

Cooperation 'lowercase c' versus transboundary water cooperation

The previous sections have explored the blurred lines, ambiguity, and overlapping nature of commonly used terms in international waters literature: water diplomacy, water governance, water management, and water cooperation. My goal is not to argue for a particular definition of diplomacy, governance, or management but rather to clarify what precisely transboundary water cooperation is or is not. A definition of each term is summarized below:

- "Water diplomacy is a dynamic and evolving process and can support important efforts to achieve *peaceful, inclusive and sustainable water cooperation* between communities, regions, and countries. It can be tailored and responsive to the uniqueness of each process and geopolitical context." (SIWI 2016)

- "Water governance refers to the range of political, social, economic and administrative systems that are in place to develop and manage water resources, and the delivery of water services, at different levels of society." (Rogers and Hall 2003, 7)
- "Transboundary water management can be defined as the application of technical tools to specific water-related questions and challenges to improve water-specific outcomes in watercourses that are transboundary by nature." (Schmeier 2018)
- "Water cooperation, defined in its broader scope, covers various levels of interactions between and among parties, stakeholders, and sectors that are involved in the development, use and management of a water resource; in the delivery of water services; or are impacted from either the actions or the consequences of such involvement. The scope covers the full cascade from local communities to transnational domain." (Adee, Aslov, and Maestu 2015, 30)

In these definitions and others, we can identify a clear point of confusion with each of these terms and cooperation. When discussing diplomacy, governance, or management, many authors reference their ability to resolve conflict and facilitate cooperation. Cooperation is inherent within each term. However, I argue that this is not transboundary water cooperation, but rather cooperation more generally – or as I call it, cooperation with a 'lowercase c.' As discussed in Chapter 4, Tuomela defines cooperation as a joint action that can be done cooperatively or non-cooperatively, depending on the attitude of the actors (Tuomela 2000; 1993). This cooperation with a 'lowercase c' is required for the existence of water diplomacy, governance, and management. For example, actors must act jointly to negotiate during a diplomatic process over shared waters while entering into the negotiations at various levels of cooperativeness (with a 'lowercase c'). Similarly, without joint action to implement management decisions, it would be difficult for states or other actors to manage shared waters on a day-to-day basis, such as with joint monitoring or infrastructure operation. Transboundary water cooperation is, then, a separate term that is related to diplomacy, governance, and management. They mutually interact, influence, and impact each other depending on the context and scale of the process and shared water resource. This is depicted by the following Figure 5.1.

Figure 5.1 shows three terms – diplomacy, governance, and management, within a Venn diagram that identifies several of the common points in the term definitions. It further identifies a few of the elements that overlap between them, with cooperation ('lowercase c') and joint action in the center – these are common themes between all three terms. Transboundary water cooperation is separated from the Venn diagram to clarify the distinction between transboundary water cooperation and the cooperation with a 'lowercase c' that is required within diplomacy, governance, and management. However, diplomacy, governance, and management can help facilitate transboundary water cooperation and resolve conflicts between actors, which is illustrated by the

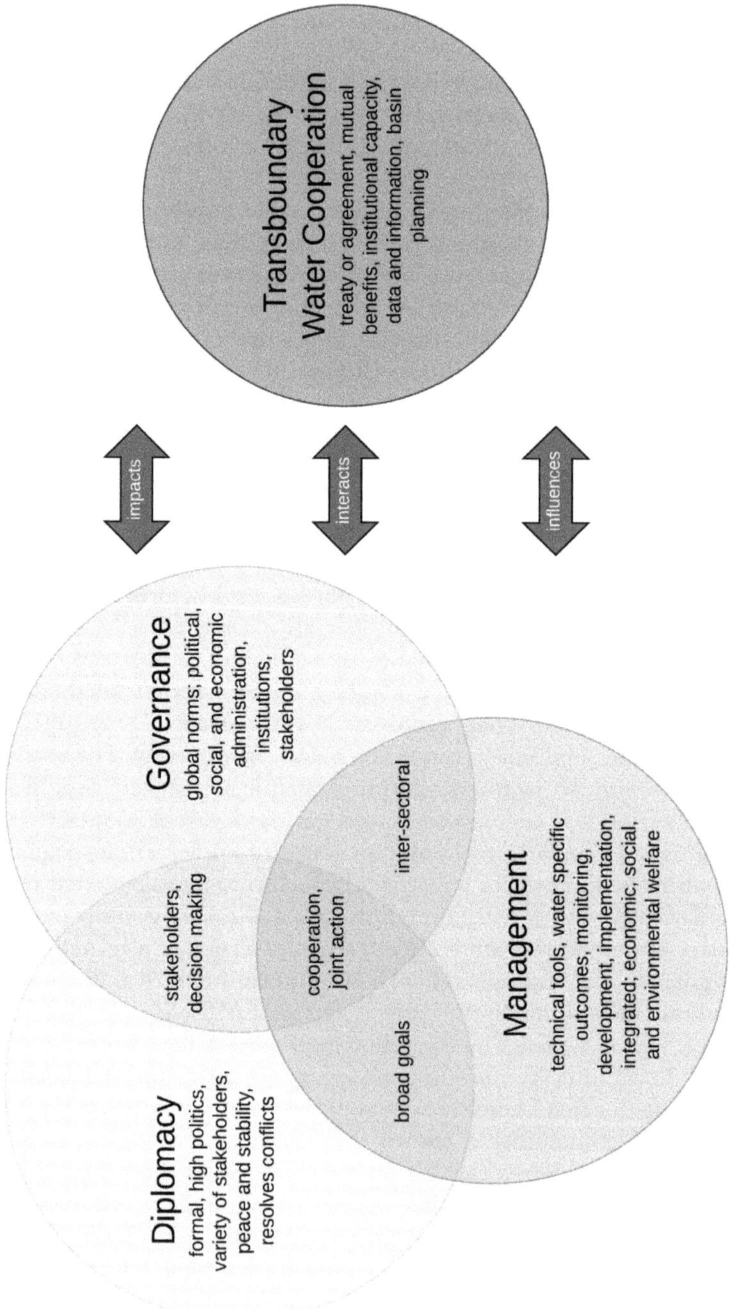

Figure 5.1 Distinction between water diplomacy, water governance, water management and transboundary water cooperation. The point of confusion is the difference between cooperation "with a lowercase c" and transboundary water cooperation.

double-ended arrows. Under transboundary water cooperation, the top five key elements mentioned by the interviewees and within the literature are listed.

A definition of cooperation

To conclude this chapter, water diplomacy, water governance, water management, and water cooperation are all amorphous and difficult to define terms, particularly since they continue to evolve with changes in the field of shared waters. Cooperation, with a 'lowercase c,' is inherent in the processes of diplomacy, governance, and management; however, this is different from the concept of transboundary water cooperation that is intertwined with these terms but separate. Transboundary water cooperation could be considered as the interactions between actors over shared waters that result in establishing mutually beneficial outcomes through a decision-making process; this process could include formal and informal legal and institutional mechanisms depending on the scale and context. Diplomacy is needed to help direct these interactions, while governance influences the norms and institutions within the decision-making process, and management can implement the mutually beneficial outcomes of this dynamic, iterative, and adaptive process. Ultimately, this chapter aims to demonstrate that developing a definition of transboundary water cooperation is not a simple task. Chapter 6 works to unscramble these complexities and develop a definition of cooperation using an explanatory framework based on place, scale, and context.

Notes

1 Please keep in mind when reading that this chapter focuses on defining *transboundary water cooperation*, not cooperation in the broader sense of the word. I use *cooperation* without the *transboundary water* modifier; however, I am referring to *transboundary water cooperation*. I will note when I refer to cooperation unlinked to transboundary water.
2 Expert interviewees included 22 Academics, 23 Practitioners, 3 Academic/Practitioners, and 6 Practitioners and Academics affiliated with NGOs: equaling a total of 54 interviews from 27 countries.
3 This breakdown is based on individual contextual units within the articles coded as having explicit mentions of definitions of cooperation. Several articles have multiple contextual units and therefore the numbers will not add up to the number of articles.

References

Adee, Zafar, Sirodjidin Aslov, and Josefina Maestu. 2015. *Water Cooperation: Views on Progress and the Way Forward*. http://proxy.library.carleton.ca/loginurl=https://www.deslibris.ca/ID/10063209.

Akhmouch, Aziza, and Francisco Nunes Correia. 2016. "The 12 OECD Principles on Water Governance – When Science Meets Policy." *Utilities Policy* 43 (December): 14–20. https://doi.org/10.1016/j.jup.2016.06.004.

Al-Saidi, Mohammad, and Amr Hefny. 2018. "Institutional Arrangements for Beneficial Regional Cooperation on Water, Energy and Food Priority Issues in the Eastern Nile Basin." *Journal of Hydrology* 562 (July): 821–31. https://doi.org/10.1016/j.jhydrol.2018.05.009.

Chaisemartin, Marguerite de. 2018. "Measuring Transboundary Water Cooperation within the Framework of Agenda 2030: A Proposal for a Revision of SDG Indicator 6.5.2." https://papers.ssrn.com/sol3/papers.cfm?abstract_id=3184575.

Chen, Sulan, John C. Pernetta, and Alfred M. Duda. 2013. "Towards a New Paradigm for Transboundary Water Governance: Implementing Regional Frameworks through Local Actions." *Ocean & Coastal Management* 85 (December): 244–56. https://doi.org/10.1016/j.ocecoaman.2012.10.019.

Chikozho, Claudious. 2014. "Pathways for Building Capacity and Ensuring Effective Transboundary Water Resources Management in Africa: Revisiting the Key Issues, Opportunities and Challenges." *Physics and Chemistry of the Earth, Parts A/B/C*, Transboundary Water Cooperation: Building Partnerships (Part 2), 76–78: 72–82. https://doi.org/10.1016/j.pce.2014.11.004.

Daoudy, Marwa, and Elizabeth J. Kistin. 2008. "Beyond Water Conflict: Evaluating the Effects of International Water Cooperation." San Francisco.

Van der Zaag, Pieter and Hubert H. G. Savenije. 2000. "Towards Improved Management of Shared River Basins: Lessons from the Maseru Conference." *Water Policy* 2 (1–2): 47–63. https://doi.org/10.1016/S1366-7017(99)00027-6.

Dinar, Ariel, Brian Blankespoor, Shlomi Dinar, and Pradeep Kurukulasuriya. 2010. "Does Precipitation and Runoff Variability Affect Treaty Cooperation between States Sharing International Bilateral Rivers?" *Ecological Economics* 69 (12): 2568–81. https://doi.org/10.1016/j.ecolecon.2010.07.036.

Finger, Matthias, Ludivine Tamiotti, and Jeremy Allouche, eds. 2006. *The Multi-Governance of Water.* New York: State University of New York Press.

Giordano, Mark, and Tushaar Shah. 2014. "From IWRM Back to Integrated Water Resources Management." *International Journal of Water Resources Development* 30 (3): 364–76. https://doi.org/10.1080/07900627.2013.851521.

Global Water Partnership. 2000. *Integrated Water Resources Management.* Stockholm: Global water partnership.

Grigg, Neil S. 2015. "Problem Archetypes as Common Ground for Water Resources Management Education." *Water Resources Management* 29 (10): 3535–50. https://doi.org/10.1007/s11269-015-1013-z.

Gupta, Joyeeta, and Claudia Pahl-Wostl. 2013. "Global Water Governance in the Context of Global and Multilevel Governance: Its Need, Form, and Challenges." *Ecology and Society* 18 (4): 53.

Ho, Selina. 2017. "China's Transboundary River Policies towards Kazakhstan: Issue-Linkages and Incentives for Cooperation." *Water International* 42 (2): 142–62. https://doi.org/10.1080/02508060.2017.1272233.

Honkonen, Tuula, and Annukka Lipponen. 2018. "Finland's Cooperation in Managing Transboundary Waters and the UNECE Principles for Effective Joint Bodies: Value for Water Diplomacy?" *Journal of Hydrology* 567 (December): 320–31. https://doi.org/10.1016/j.jhydrol.2018.09.062.

Huntjens, Patrick, Yumiko Yasuda, Ashok Swain, Rens de Man, Bjørn-Oliver Magsig, and Shafiqul Islam. 2016. "The Multi-Track Water Diplomacy Framework: A Legel and Political Economy Analysis for Advancing Cooperation over Shared Waters."

Hague: The Hague Institute for Global Justice. www.thehagueinstituteforglobaljustice.org/thinking-doing/publication/the-multi-track-water-diplomacy-framework-a-legal-and-political-economy-analysis-for-advancing-cooperation-over-shared-waters/.

International Conference on Water and the Environment. 1992. "The Dublin Statement." January 31, 1992. www.wmo.int/pages/prog/hwrp/documents/english/icwedece.html.

Jägerskog, Anders. n.d. "Transboundary Water Management – Why It Is Important and Why It Needs to Be Developed." Stockholm International Water Institute and UNDP-Shared Waters Partnership.

Keohane, Robert O. 1984. *After Hegemony: Cooperation and Discord in the World Political Economy*. Princeton: Princeton University Press. https://academic.oup.com/ia/article-lookup/doi/10.2307/2617490.

———. 1989. *International Institutions and State Power: Essays in International Relations Theory*. Boulder: Westview Press.

Keskinen, Marko, Antti Inkinen, Ulla Hakanen, Antti Rautavaara, and Markku Niinioja. 2013. "Water Diplomacy – Bringing Diplomacy into Water Cooperation and Water into Diplomacy." In *Hydro-Diplomacy: Sharing Waters Across Borders*, edited by Ganesh Pangare. New Delhi, India: Academic Foundation. 35–40.

Klimes, Martina, David Michel, Elizabeth Yaari, and Phillia Restiani. 2019. "Water Diplomacy: The Intersect of Science, Policy, and Practice." *Journal of Hydrology*, March. https://doi.org/10.1016/j.jhydrol.2019.02.049.

Milly, Paul C.D., Julio Betancourt, Malin Falkenmark, Robert M. Hirsch, Zbigniew W. Kundzewicz, Dennis P. Lettenmaier, and Ronald J. Stouffer. 2008. "Stationarity Is Dead: Whither Water Management?" *Science* 319 (February): 573–74.

Molnar, Kata, Rosa Cuppari, Susanne Schmeier, and Siegfried Demuth. 2017. "Preventing Conflicts, Fostering Cooperation – The Many Roles of Water Diplomacy." Stockholm: UNESCO's International Centre for Water Cooperation.

Mostert, Erik, Eelco van Beek, N.W.M Bouman, E. Hey, Hubert H. G. Savenije, and W.A.H Thissen. 2000. "River Basin Management and Planning." In *River Basin Management*, 24–55. Proceedings of the International Workshop on River Basin Management on 27–29 October 1999, No 31. The Hague: UNESCO-IHP.

Mumme, Stephen. 1993. "Innovation and Reform in Transboundary Resource Management: A Critical Look at the International Boundary and Water Commission, United States and Mexico." *Natural Resources Journal* 33 (1): 93–120.

OECD. 2015. "OECD Principles on Water Governance." Paris: Organisation for Economic Co-operation and Development. www.oecd.org/cfe/regional-policy/OECD-Principles-on-Water-Governance-brochure.pdf.

Pahl-Wostl, Claudia, Paul Jeffrey, and Jan Sendzimir. 2011. "Adaptive and Integrated Management of Water Resources." In *Water Resources Planning and Management*, edited by R. Quentin Grafton and Karen Hussey, 61–73. Cambridge: Cambridge University Press. http://link.springer.com/10.1007/978-3-540-75941-6_4.

Pahl-Wostl, Claudia, P. Kabat, and Jörn Möltgen, eds. 2008. *Adaptive and Integrated Water Management: Coping with Complexity and Uncertainty*. Berlin; New York: Springer.

Pahl-Wostl, Claudia, Louis Lebel, Christian Knieper, and Elena Nikitina. 2012. "From Applying Panaceas to Mastering Complexity: Toward Adaptive Water Governance in River Basins." *Environmental Science & Policy* 23 (November): 24–34. https://doi.org/10.1016/j.envsci.2012.07.014.

Phillips, David, Marwa Daoudy, Stephen McCaffrey, Joakim Öjendal, and Anthony Turton. 2006. *Trans-Boundary Water Cooperation as a Tool for Conflict Prevention and Broader Benefit-Sharing*. Ministry for Foreign Affairs Stockholm.

Pohl, Benjamin, Ken Conca, Geoffrey D. Dabelko, Annika Kramer, David Michel, Susanne Schmeier, Ashok Swain, and Aaron T. Wolf. 2014. "The Rise of Hydro-Diplomacy." Climate Diplomacy. Berlin: Adelphi and Federal Foreign Office.

Rahaman, Muhammad Mizanur. 2009. "Principles of Transboundary Water Resources Management and Ganges Treaties: An Analysis." *International Journal of Water Resources Development* 25 (1): 159–73. https://doi.org/10.1080/07900620802517574.

Rogers, Peter, and Alan W Hall. 2003. *Effective Water Governance*. Technical Committee Background Papers 7. Stockholm: Global Water Partnership.

Sadoff, Claudia W., and David Grey. 2005. "Cooperation on International Rivers." *Water International* 30 (4): 420–27. https://doi.org/10.1080/02508060508691886.

Salamé, Léna. 2018. Research InterviewSkype.

Schmeier, Susanne. 2016. "Water Diplomacy at the Basin-Level." Presented at the International Symposium on Water Diplomacy, Stockholm, November. www.siwi.org/wp-content/uploads/2016/12/Susanne-Schmeier-presentation.pdf.

———. 2018. "What Is Water Diplomacy and Why Should You Care?" *Global Water Forum* (blog). August 31, 2018. www.globalwaterforum.org/2018/08/31/what-is-water-diplomacy-and-why-should-you-care/.

Schmeier, Susanne, Andrea K. Gerlak, and Sabine Blumstein. 2016. "Clearing the Muddy Waters of Shared Watercourses Governance: Conceptualizing International River Basin Organizations." *International Environmental Agreements: Politics, Law and Economics* 16 (4): 597–619. https://doi.org/10.1007/s10784-015-9287-4.

Schmeier, Susanne, and Birgit Vogel. 2018. "Ensuring Long-Term Cooperation Over Transboundary Water Resources Through Joint River Basin Management." In *Riverine Ecosystem Management*, edited by Stefan Schmutz and Jan Sendzimir, 347–70. Cham: Springer International Publishing. https://doi.org/10.1007/978-3-319-73250-3_18.

Schroeck, Nicholas. 2018. Research Interview.

Selby, Jan. 2013. "Cooperation, Domination and Colonisation: The Israeli-Palestinian Joint Water Committee." *Water Alternatives* 6 (1): 24.

Shmueli, Deborah F. 1999. "Water Quality in International River Basins." *Political Geography* 18 (4): 437–76. https://doi.org/10.1016/S0962-6298(98)00106-1.

SIWI. 2016. "International Symposium on Water Diplomacy: An Overview." *Stockholm International Water Institute* (blog). 2016. www.siwi.org/international-symposium-water-diplomacy-overview/.

Strategic Foresight Group. 2015. *Water Cooperation Quotient*. Mumbai, India: Strategic Foresight Group. www.strategicforesight.com/publication_pdf/28799WCQ-web.pdf.

Svendsen, Mark (Ed). 2005. *Irrigation and River Basin Management: Options for Governance and Institutions*. CABI; International Water Management Institute (IWMI). https://cgspace.cgiar.org/handle/10568/36480.

Tuomela, Raimo. 1993. "What Is Cooperation?" *Erkenntnis* 38 (1): 87–101. https://doi.org/10.1007/BF01129023.

———. 2000. *Cooperation*. Philosophical Studies Series 82. Dordrecht, Netherlands: Kluwer Academic Publishers.

Turton, Anthony R., and Anton Earle. 2005. "Institutions in South African International River Basins." In *Water Institutions: Policies, Performance and Prospects*, edited by Chennat

Gopalakrishnan, Asit K. Biswas, and Cecilia Tortajada, 154–73. Water Resources Development and Management. Berlin, Heidelberg: Springer Berlin Heidelberg. https://doi.org/10.1007/3-540-26567-8_7.

United Nations. 1997. *Convention on the Law of Non-Navigational Uses of International Watercourses. A/RES/51/229.* http://heinonline.org/hol-cgi-bin/get_pdf.cgi?handle=hein.journals/tilj33&section=22.

United Nations Economic Commission for Europe. 2013. 1992 *Convention on the Protection and Use of Transboundary Watercourses and International Lakes Amended.* www.unece.org/fileadmin/DAM/env/documents/2013/wat/ECE_MP.WAT_41.pdf.

Villholth, Karen G., Elena López-Gunn, Kirstin Conti, Alberto Garrido, and Jac A. M. van der Gun, eds. 2018. *Advances in Groundwater Governance.* Leiden, The Netherlands: CRC Press/Balkema.

"Water Diplomacy@Tufts." 2019. 2019. http://ase.tufts.edu/igert/waterDiplomacy/documents/keyIdeasPrinciples.pdf.

West, Stuart A., Ashleigh S. Griffin, and Andy Gardner. 2007. "Evolutionary Explanations for Cooperation." *Current Biology* 17 (16): R661–72. https://doi.org/10.1016/j.cub.2007.06.004.

Wiegleb, Viviana, and Antje Bruns. 2018. "Hydro-Social Arrangements and Paradigmatic Change in Water Governance: An Analysis of the Sustainable Development Goals (SDGs)." *Sustainability Science* 13 (4): 1155–66. https://doi.org/10.1007/s11625-017-0518-1.

Wolf, Aaron T. 2007. "Shared Waters: Conflict and Cooperation." *Annual Review of Environment and Resources* 32 (1): 241–69. https://doi.org/10.1146/annurev.energy.32.041006.101434.

———. 2010. *Sharing Water, Sharing Benefits: Working towards Effective Transboundary Water Resources Management. A Graduate/Professional Skills-Building Workbook.* UNESCO: International Hydrological Programme: Oregon State University. http://unesdoc.unesco.org/images/0018/001893/189394e.pdf.

Wolf, Aaron T., Shira B. Yoffe, and Mark Giordano. 2003. "International Waters: Identifying Basins at Risk." *Water Policy* 5 (1): 29–60. https://doi.org/10.2166/wp.2003.0002.

World Bank. 1989. "From Crisis to Sustainable Growth – Sub Saharan Africa: A Long-Term Perspective Study." 8209. The World Bank. http://documents.worldbank.org/curated/en/498241468742846138/From-crisis-to-sustainable-growth-sub-Saharan-Africa-a-long-term-perspective-study.

Yasuda, Yumiko, Douglas Hill, Dipankar Aich, Patrick Huntjens, and Ashok Swain. 2018. "Multi-Track Water Diplomacy: Current and Potential Future Cooperation over the Brahmaputra River Basin." *Water International* 43 (5): 642–64. https://doi.org/10.1080/02508060.2018.1503446.

Yeophantong, Pichamon. 2017. "River Activism, Policy Entrepreneurship and Transboundary Water Disputes in Asia." *Water International* 42 (2): 163–86. https://doi.org/10.1080/02508060.2017.1279041.

Zartman, I. William, ed. 2008. "Concept: Cooperation." *PIN Points* 30: 5–7.

6 Framing transboundary water cooperation

A framework definition

This is the final chapter of Part II, which has been answering the question: How is transboundary water cooperation defined? The previous chapter identified the type of definitions used to discuss cooperation[1] (see Table 5.1) and presented the most commonly referenced key elements of cooperation. Key elements of cooperation are words or phrases representing a standalone concept or component of cooperation; 50 key elements were identified in total (see Appendix A for complete list). Building on the data analysis from the qualitative content analysis of expert[2] interviews, peer-reviewed articles, and grey literature, I use a frame analysis to make sense of the wide range of key elements of cooperation identified. Frames are a tool to organize various perspectives on a topic (Entman 1993; Goffman 1974), such as cooperation, which has many perspectives on what constitutes it. From this frame analysis, this chapter develops the Four Frames of Transboundary Water Cooperation, which is a framework for understanding and 'defining' transboundary water cooperation that can be consistently applied despite variations in scale, context, and place and the ultimate goal of Part II.

Introducing the Four Frames of Cooperation

To define cooperation, we need to understand the core components – or key elements – that constitute the concept. Chapter 5 discusses the results of the qualitative content analysis of 54 expert interviews and the literature sample (n = 273); however, as expected from the diversity of definitions, there is a wide range of key elements present. While many, as highlighted in Table 5.2, are commonly referenced, other key elements are not. The question then becomes: How can these 50 distinct key elements be used as a foundation to create a definition of transboundary water cooperation that is comprehensive, clear, and consistent?

Using mind mapping as a tool to conduct a frame analysis, it is apparent that the key elements and their sub-themes fall into Four Frames of Cooperation. I categorized them by grouping key elements together into sub-themes and grouping them until the four frames were apparent in the data; this was done using inductive reasoning. On the following page, the mind map in Figure 6.1

DOI: 10.4324/9781003126249-8

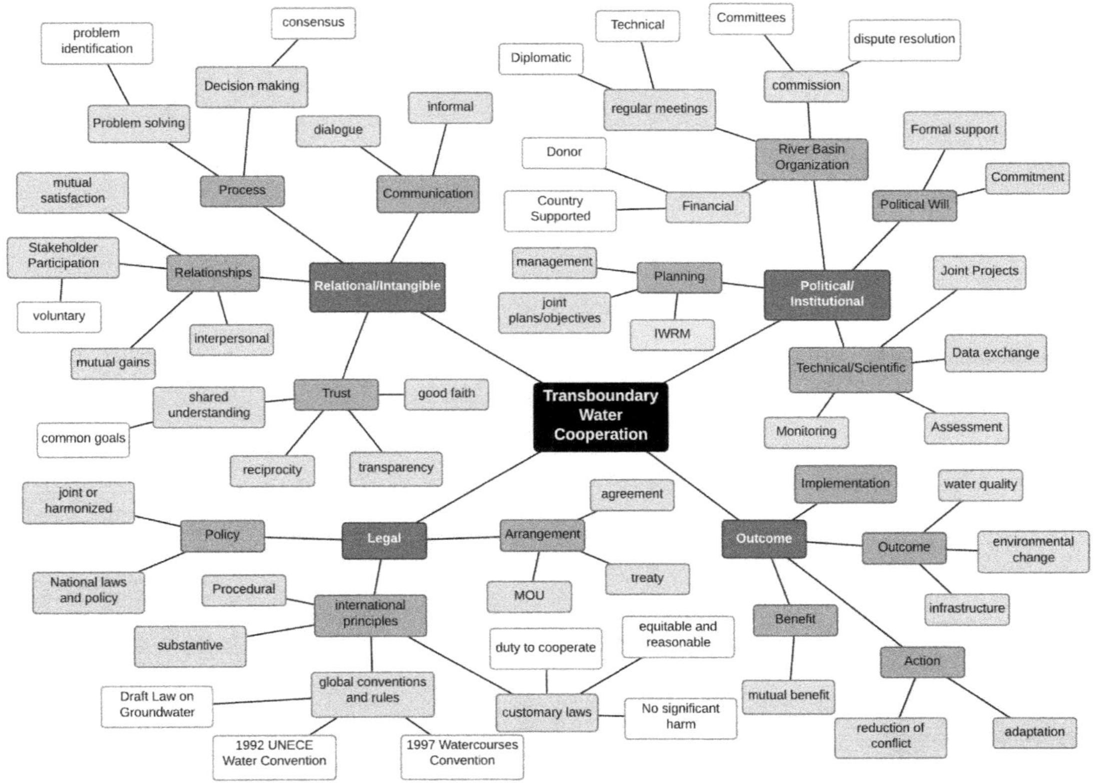

Figure 6.1 Mind map of the Four Frames of Cooperation.

This illustrates how the key elements were grouped into themes, which resulted in the Four Frames of Cooperation.

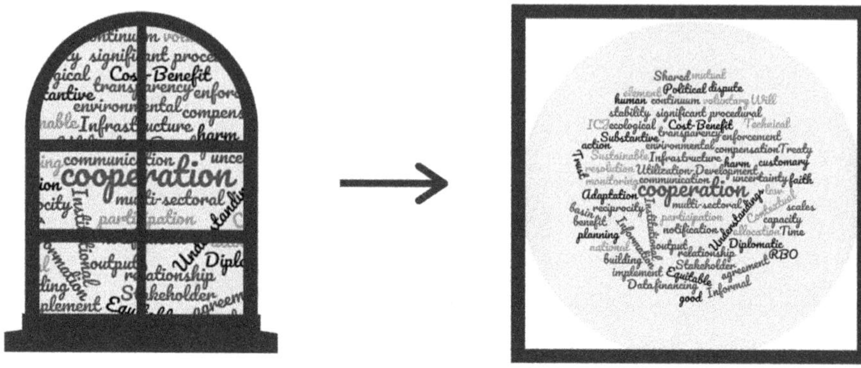

Figure 6.2 Graphic analogy of how frames alter perspective.

The figure on the left represents the window frame that restricts or defines the view. The figure on the right represents the room. A window is on each of the room's four walls, which contains transboundary water cooperation depicted as a word cloud.

illustrates how key elements fit within the four frames: legal, institutional, outcomes, and relational.

The interviewees and the literature tended to discuss cooperation from specific perspectives, often noting multiple frames within a definition or discussion, yet very few mentioned all four frames together. These frames represent a view or perspective into transboundary water cooperation, much like a window that frames a view into a room. If we consider the definitions of transboundary water cooperation as being contained within a four-sided room, each frame of cooperation would be a window looking into that room (see Figure 6.2). When looking into a room from outside, you can see portions of the room from multiple windows, but each view is different.

For example, when looking into a living room from one window, a couch and a chair are visible, yet the chair and a T.V. can be seen from another window. This is the same with the Four Frames of Cooperation – elements are visible and may overlap between frames, but they are viewed differently depending on the perspective you view them through. Therefore, to develop a comprehensive view of transboundary water cooperation, we must recognize our biased perspectives on cooperation and incorporate the other perspectives into the definition. The Four Frames of Transboundary Water Cooperation are legal, institutional, relational, and outcome.

The legal frame of cooperation

Of the Four Frames of Transboundary Water Cooperation, the legal frame of cooperation is one of the most recognizable and established. Legal definitions of cooperation are arguably the most easily measured, for example, the existence of a treaty, which lends to their prevalence in the literature. However, when combining references from literature and interviews, as well as from implicit and explicit definitions of cooperation, it is the third most referenced frame of the four in the data, coming after the institutional and relational frames. In the interviews, legal elements of cooperation were cited the second most (tied with institutional elements) after relational elements of cooperation. The following key elements were categorized as belonging to the legal frame: treaty or agreement, procedural, dispute resolution, substantive elements, national laws, notification, equitable, significant harm, enforcement, customary law, and compensation. The most mentioned key element by far was a reference to a treaty or agreement, which is one of the most common ways to measure cooperation or a component of a method to measure cooperation in the literature (see for examples Ide and Detges 2018; Dinar et al. 2010; Wolf, Yoffe, and Giordano 2003; De Stefano et al. 2017; Bernauer and Siegfried 2008; Brochmann 2012; TFDD 2021). Procedural and substantive elements were also commonly referenced; these included references to procedural and substantive legal elements generally or specific procedural or substantive legal mechanisms that were not the most commonly cited, for example, regular meetings. These were coded separately, such as equitable and reasonable use, no significant harm, dispute resolution, notification, enforcement, and compensation. For example, "substantive rules alone are insufficient to ensure the effective implementation of international agreements. Operational rules of procedure and institutional mechanisms have an integral role to play" (Su 2014) is coded as procedural. Whereas, "most importantly for this article, however, was concern surrounding the relationship between the principles of equitable and reasonable use and the duty not to cause significant harm, which together provide the foundation for a cooperative framework on transboundary waters" (Devlaeminck 2018) is coded both as equitable and significant harm. The following word cloud, Figure 6.3, illustrates the most common words used in the implicit and explicit discussions of cooperation coded as within the legal frame given the key elements noted in the definitions. This cloud includes terms that overlap with other frames of cooperation.

So, what do legally framed definitions of cooperation look like? Of the direct, explicit definitions, 10 of the interviews and 15 literature definitions were coded within the legal frame. The following are two examples from both datasets:

- Interview: "I'm biased because I come from a legal perspective, so when I think of transboundary water cooperation, I think of it mainly from an international legal perspective. That would mean that when you have transboundary water cooperation, there is some sort of international legal instrument or arrangement that allows riparian states, be it surface or groundwater, to develop jointly some sort of management of that specific water resource." (Sindico 2018)
- Literature: "Cooperation, when defined as a dependent variable in causal explanations of international water management, is usually measured in

Figure 6.3 Legal frame word cloud.

> binary terms. That is, with a yes/no answer to the questions whether an agreement, treaty or international institution is in place." (Bernauer and Siegfried 2008)

Generally, there are two categories of legal frames in terms of cooperation: (1) the existence of a treaty or agreement and (2) provisions in line with international water law. These categories understand transboundary water cooperation as the existence of a treaty or signing an additional agreement. Or, they understand cooperation as arrangements that adhere to various articles within the 1997 U.N. Watercourses Convention, the 1992 UNECE Water Convention, or customary law. From this, we can develop a definition for the perspective of transboundary water cooperation captured by this legal framing. Overall, legal frames of transboundary water cooperation view cooperation as formal legal elements, including the existence of a treaty or agreement, adherence to conventions, and the inclusion of key substantive and procedural principles.

The institutional frame of cooperation

Definitions of transboundary water cooperation that fall into the institutional frame are the most commonly referenced in the data. However, we see a difference between the implicit discussions, which reference institutional frames most, while the explicit definitions reference relational frames most, followed by institutional frames. In the interview data, institutional frames are the second most commonly referenced (tied with legal frames) after relational frames. I would argue that the prevalence of institutional frames as a perspective on cooperation is due to the landmark studies on institutional capacity and the role of river basin organizations. The Basins at Risk project demonstrated that institutional capacity is critical to a basin's ability to react to change and prevent conflict (Wolf, Yoffe, and Giordano 2003). Later, Schmeier and Blumstein took this further by considering the role of river basin organizations in governing watercourses (Schmeier 2013; Blumstein and Schmeier 2017; Schmeier 2015). Furthermore, the idea of institutions supporting cooperation has a strong basis outside of international waters literature, with international relations scholars promoting institutions as the means to promote cooperation and order in the international system (Zartman 2008; Keohane 1984; Herbert 1996). Ostrom and others have also done extensive work on the role of institutions in managing common-pool resources (Ostrom 2010; 1990; Vollan and Ostrom 2010; Ostrom 2005; 2008). Ultimately, the existence and scope of the functions of institutions are easily measured, much like the existence of legal arrangements, when developing methods to measure the extent of transboundary water cooperation (see for examples De Stefano et al. 2017; 2010; TFDD 2021).

The key elements that were categorized as part of the institutional frame include institutional capacity, basin planning, technical cooperation, river basin organizations, human capacity, financing, diplomatic cooperation, and data and information. The most commonly referenced element is institutional capacity. Generally, text within the institutional capacity code references the term in general and does not specify what is meant by institutional capacity. However, I also include text that references the combination of agreements and river basin organizations or discusses institutions more generally. Basin planning, river basin organizations, technical cooperation, and data and information were also commonly referenced in the data. Text within the data and information code is categorized as part of the institutional frame rather than the legal frame. Generally, data and information were discussed in the practical sense of collecting, monitoring, and sharing data as part of an institution's function. Discussions of legal requirements for data and information exchange were less frequent and included in the procedural code mentioned in the legal frame. This example of data and information illustrates the overlap between the frames of cooperation and how data and information can be viewed differently

Figure 6.4 Institutional frame word cloud.

depending on the frame of reference. The following word cloud, Figure 6.4, illustrates the most common words used in the implicit and explicit discussions of cooperation within the institutional frame. As with the other word clouds, this cloud includes terms that overlap with other frames of cooperation.

So, what do definitions of cooperation using an institutional frame look like in the data? There are 10 definitions in the interviews and 15 sources with explicit definitions in the literature. The following are two examples from both datasets:

- Interviews: "I think the first element about transboundary cooperation is, I would say, dialogue. Number one is dialogue. Dialogue means that you have platforms, you have institutions, you have the capacities, which really initiate this and keeps the dialogue." (Abdullaev 2018)

- Literature: "Essential structural features of cooperation include: government participation, principal and subsidiary organs, voting procedure, and functions and powers of the agency." (Caponera 1985)

The interview definition also presents an example of an overlap between the institutional and relational frames of cooperation; it acknowledges that not only are platforms and institutions needed but that dialogue – a relational aspect of cooperation – is also needed. From this analysis and these data, we can develop a definition for the institutional perspective on transboundary water cooperation. Institutional cooperation is defined as having formal institutions, such as river basin organizations, that support the development and implementation of joint basin planning and other activities that are needed to manage shared water resources. In addition, it captures the political nature of these institutions and the will of stakeholders to support their operation, function, and scope.

The relational frame of cooperation

The third frame of transboundary water cooperation is relational. This is the second most commonly referenced frame overall in the data and the third most referenced in the literature alone. However, definitions coded as relational were the most common in the data from the interviews. While this research was not designed to determine why these differences exist, I would propose that the prevalence of legal and institutional frames as the most common definitions in the literature is due to their measurability, particularly because measuring the relational elements of cooperation is quite difficult. We can see a similar distribution of definitions between the relational, institutional, and legal frames between the interviewees of different backgrounds. Academics referenced the legal frame most often, although the margin between counts was only one or two definitions between the four frames. While practitioners strongly favored definitions in the relational and institutional frames, they also made references to the outcome and legal frames. While I did not collect data on why these differences exist between academics and practitioners, I would speculate that the relational and institutional aspects of cooperation are the most relevant for the day-to-day operations of cooperating over shared waters. Academics, on the other hand, may focus on a greater range of scales and contexts and therefore reference all four frames.

The key elements categorized as a part of the relational frame are communication, participation, shared understandings, mutual benefit, trust, reciprocity, good faith, transparency, relationship building, process, and informal. The most commonly referenced key element in the relational frame is mutual benefit. The text coded in the mutual benefit code is the exact phrase 'mutual benefit,' mutual satisfaction, or text that refers to outcomes of cooperation that meet the needs of multiple parties in an equitable manner. This key element – mutual benefit – overlaps with the outcome frame. Mutual benefits fall within the relational frame because the concept of *mutually shared* benefits embodies the nature of the relationship between actors. Where benefits can be unilateral or inequitably distributed, it is the shared and equitable nature of 'mutual' that is captured by the relational frame and 'benefits' is captured by the outcomes frame. The other most commonly referenced key elements are trust, shared understandings, and communication. An additional key element to note in the relational frame is process. Process includes decision making, problem identification, and problem solving when discussed in a general or non-specific manner. It does not include legal or institutional procedural mechanisms. Process is included in this frame because legal and institutional features could be in place, but the process of cooperation is not functioning or not effective. In this frame, process embodies the relationships required to initiate, develop, implement, and maintain cooperation. These process relational mechanisms can accompany the legal and institutional structure or aid in the development of informal cooperation, depending on the scale and context.

The key elements of the relational frame attempt to capture the interpersonal nature of cooperation, as well as the scaled-up relationships between state actors. This frame is based on the idea of cooperative attitudes through which joint action is taken in Tuomela's representation of cooperation (Tuomela 1993). Relational cooperation and its core concepts of trust, relationship building, communication, and so on are challenging aspects to measure, particularly on a large scale. The following word cloud, Figure 6.5, illustrates the most common words used in the implicit and explicit discussions of cooperation within the relational frame. As with the other word clouds, this cloud includes terms that overlap with other frames of cooperation.

So, what do definitions of cooperation using a relational frame look like in the data? There are 20 sources in the interviews and 14 sources in the literature that include relational aspects of cooperation in their definitions. The following are two examples from both datasets:

- Interviews: "It's recognizing that the river doesn't start and end in Swaziland, for instance. … and then the cooperation element of it, while I am using the river in Swaziland, I should be cognizant of the needs of Mozambique and while South Africa is using the river before me, they should leave enough water to know that I also need the same water. So, the cooperation is on working together in that context of that river, which is basically ours to share, but we all try to satisfy our needs and work in a context where we all try to make sure each one gets a bit." (Mthimkhulu 2017)
- Literature: "Water cooperation refers to the peaceful management and use of freshwater resources at local, national, regional and international levels among various players and sectors. The concept of water cooperation entails working together towards a common goal, in a way that is mutually beneficial." (International Annual UN-Water Zaragoza Conference 2012/2013, 2017 in Molnar et al. 2017)

The interview definition is a clear example of relational aspects of cooperation from a practitioner that demonstrates shared understandings and values that

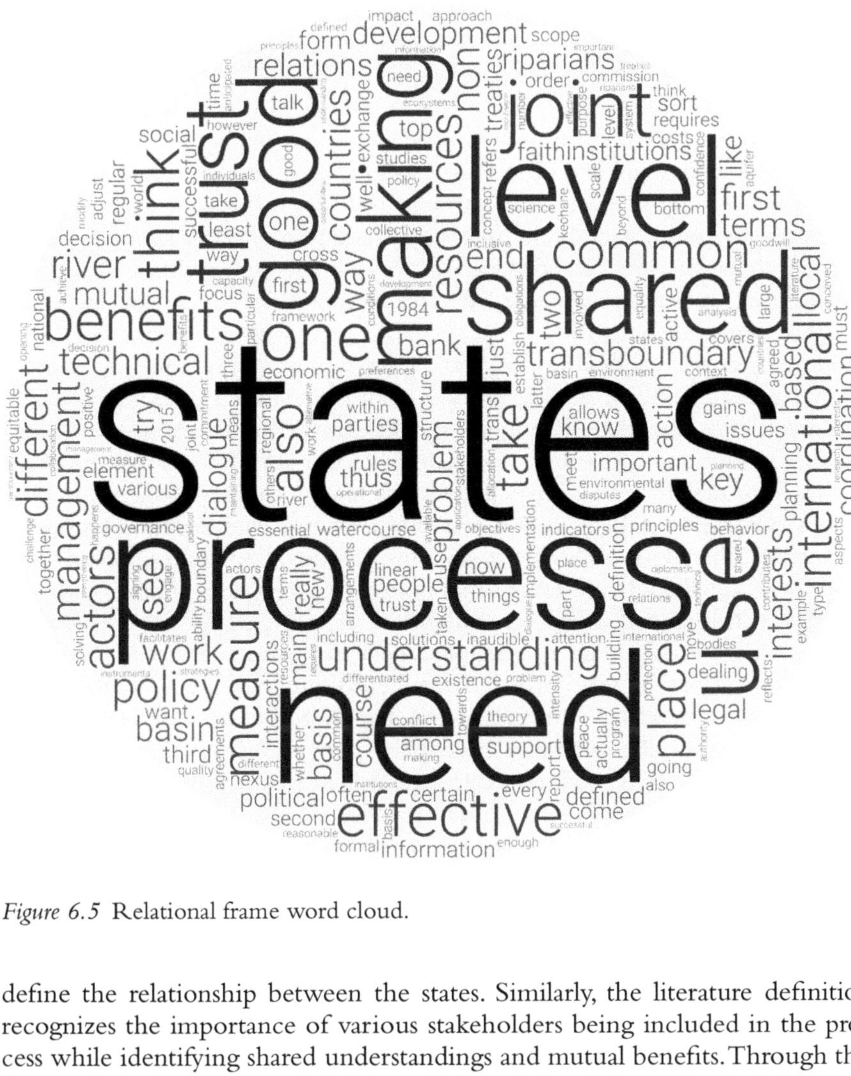

Figure 6.5 Relational frame word cloud.

define the relationship between the states. Similarly, the literature definition recognizes the importance of various stakeholders being included in the process while identifying shared understandings and mutual benefits. Through this data analysis, I propose a definition of the relational frame for definitions of transboundary water cooperation: relational cooperation addresses the process and relationship building between all relevant stakeholders, including trust and shared understandings, while working transparently with communication to identify and pursue mutually beneficial outcomes.

The outcome frame of cooperation

The final frame is the least referenced in the data, that is, outcome cooperation. This is the least referenced in both literature and interviews, as well as by both academics and practitioners. Much of the literature on cooperation does not define the term 'outcome,' which is the gap addressed by this chapter, nor do they offer metrics to measure the outcomes of cooperation (Mianabadi et al. 2015). Furthermore, measures of cooperation most often evaluate a formal process and do not systematically assess the outcomes of that process (Earle and Wouters 2015). However, some definitions of cooperation, within the international water literature and other fields, include an outcome component, such as joint action, mutual goal, fitness benefit, and so on. If cooperation is a process to achieve an intended outcome, then the frame of outcome cooperation attempts to capture the intended benefits of cooperative processes. This frame is particularly relevant to defining effective cooperation, which is addressed in the next chapter.

The key elements identified as part of the outcome frame include mutual benefit, reduce conflict, ecological and environmental change, implementation, adaptation, cost-benefit, outputs, stability, infrastructure, sustainability, and use and development. The most common code is mutual benefits. As discussed in the relational frame, 'mutual benefits' is an overlapping code. Outcome cooperation captures the benefit aspect, while relational cooperation embodies the mutual, shared, or equitable distribution of those outcomes. Benefits are not inherently shared, and actors may enter a cooperative process with the sole desire to achieve their own outcomes or to free ride on the process. Hence, while outcomes are an essential aspect and often a reason for actors to enter cooperation, the relational element is equally important. Mutual benefits are often mentioned in definitions, yet what those benefits are is often not defined. A few specific outcomes are explicitly mentioned in the data and are coded as key elements, such as reducing conflict, sustainability, and environmental change; in addition, economic benefits are also included. Reduction of conflict is different from the legal code of dispute resolution, as this is the goal of cooperation – to reduce, prevent, or resolve conflict, compared to the legal mechanisms of dispute resolution. For example, Dr. Andrea Gerlak of the University of Arizona stated that cooperation is "the absence of a lot of conflict. It doesn't look like everyone's happy, or that people are doing great things together, like building things or collecting data together. It is oftentimes expressed or can be seen as the absence of a lot of conflict" (Gerlak 2018). These key elements and other outcomes of cooperation are dependent on context and scale and will vary between basins and aquifers. The following word cloud, Figure 6.6, illustrates the most common words used in the implicit and explicit discussions of cooperation within the outcome frame. As with the other word clouds, this cloud includes terms that overlap with other frames of cooperation.

So, how do definitions within the outcome frame appear? Within the data, there are nine sources in the literature that include explicit, direct definitions incorporating outcome framing and five sources in the interviews. The following are two examples of outcome framed definitions of transboundary water cooperation from both datasets:

Figure 6.6 Outcome frame word cloud.

- Interviews: "Well it incorporates at least two parties, maybe more, on an equal basis. Of course, the structures behind the actual cooperation, they may be many and in different parts of the world and different basins as well. But, it's mutual cooperation based on trust and based on, hopefully, consensus so that one can agree upon about the benefits and these outcomes together." (Sallisalmi 2018)
- Literature: "The main aim of transboundary water cooperation, apart from crisis and conflict prevention, is poverty reduction and resource protection." (National Federal Ministry for Economic Cooperation and Development in Molnar et al. 2017)

The literature quote defines cooperation based on its aims or outcomes, while the interview definition incorporates multiple frames. The idea of mutual cooperation, consensus, and equal standing all stem from the frame of relational cooperation, while the benefits and outcomes come from the outcome frame of cooperation. From this data analysis, I propose that the outcome frame of cooperation be defined as the benefits, goals, actions, or specific outputs of cooperation. This can include inequitable, unilateral, or mutual benefits, as defined by the parties or by international water law and norms, achieved through joint action.

The Four Frames of Cooperation framework

These four frames – legal, institutional, relational, and outcome – are the core aspects needed for full transboundary water cooperation. Therefore, I have incorporated the ideas behind these frames into a general definition of transboundary water cooperation. I propose that transboundary water cooperation be defined as *interactions between actors over shared waters that result in establishing mutually beneficial outcomes through a decision-making process; this process could include formal and informal legal and institutional mechanisms depending on the scale and context.*

Like most general definitions, this definition is broad and relatively generic, which makes it applicable to many situations, but this is also a limitation. Hence, this definition is meant to accompany the Four Frames of Transboundary Water Cooperation, which I call the Four Frames of Cooperation for ease of comprehension and reading. This framework aims to provide a greater depth and nuance to the definition that incorporates the four frames. The legal and institutional frames are quite clear in the definition as the formal and informal structures in place to further cooperation. The relational frame is embodied through the idea of mutually shared benefits as well as the decision-making process. The outcome frame is also captured in the idea of beneficial outcomes. Each of these frames can be identified within a cooperative process or perhaps as missing from a cooperative process. The key elements of each frame can help users identify what aspects of their cooperative processes fit within each frame. See the following diagram, Figure 6.7, for a visual representation of the Four Frames of Cooperation framework. In this diagram, the four frames are defined as components of transboundary water cooperation, with example key elements listed for each frame. What mechanisms, relations, or outcomes are needed or exist in a particular basin or aquifer depend on scale and context; the key elements provided are examples to help the user further understand what each frame represents.

With its rise in the international agenda, transboundary water cooperation is being promoted as a means to improve water management and sustainable development. However, as several researchers have shown, particularly the London Water Research Group, not all cooperation can be qualified as good, nor all conflict as bad (see Zeitoun and Mirumachi 2008; Zeitoun, Mirumachi, and Warner 2011; Zeitoun et al. 2017; Mirumachi and Allan 2007; Mirumachi 2007). The dichotomy between conflict and cooperation perpetuates this view

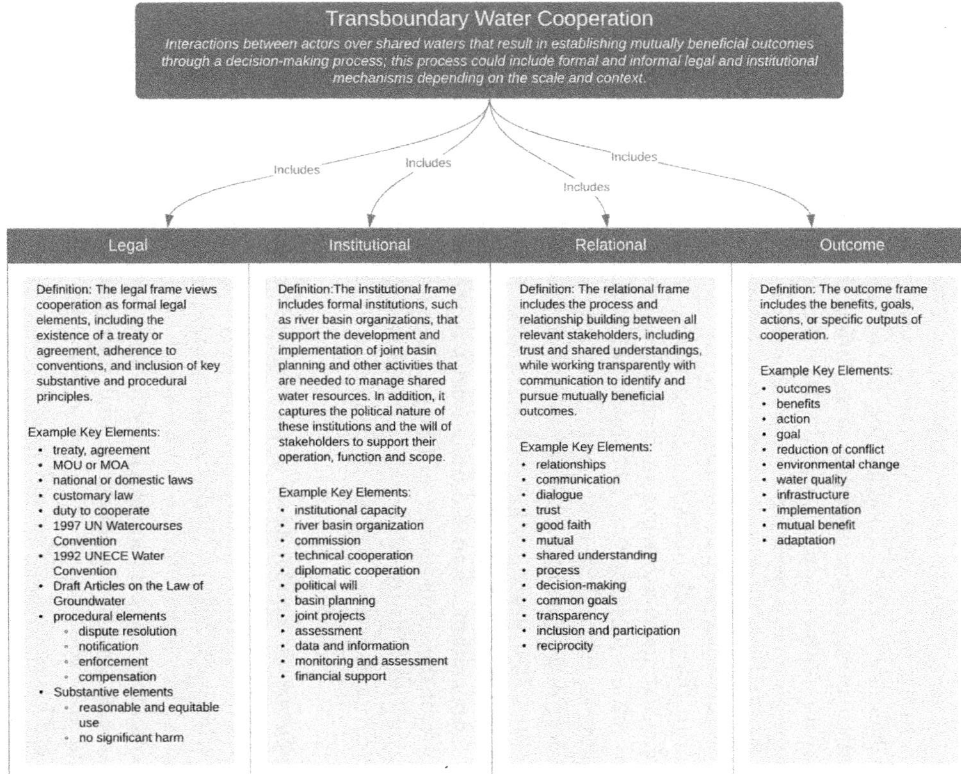

Figure 6.7 Initial Four Frames of Cooperation.

This flowchart depicts the Four Frames of Cooperation and provides examples of key elements within each frame.

that cooperation in all forms is positive (Zeitoun and Mirumachi 2008). This oversimplification of transboundary water cooperation ignores the potential for inequitable, unsustainable, or ineffective cooperation and the potential for conflict to be constructive in developing future collaboration rather than maintaining an inequitable status quo. Therefore, the definition of transboundary water cooperation presented in this framework is neutral. Depending on the basin or aquifer and the scale and context, the Four Frames of Cooperation can be interpreted as positively or negatively contributing to equitable and sustainable transboundary water cooperation. For example, there may be a lack of trust or a strong power differential creating an unequal relationship between riparian stakeholders that is contributing toward the negative interpretation of cooperation in the relational frame. This differs from many other existing measures of the cooperation, such as the Blue Peace Index, which considers "water cooperation as a 'positive-sum' outcome in which participants enjoy more benefits from working together than they would working alone" (The Economist Intelligence Unit 2020, 15).

Instead, this framework holds that cooperation is a neutral term, and it is a process that can be constructive or destructive, as defined by Zeitoun et al. (2020). The authors identify types of cooperation that fall within the categories of constructive or destructive cooperation. Constructive cooperation promotes and aids in reaching an equitable arrangement, whereas destructive cooperation includes actions that move cooperation away from being equitable (Zeitoun, Mirumachi, and Warner 2020). These types of cooperation fit within the Four Frames of Cooperation, given the framework's neutrality and the potential to evaluate each of the Four Frames of Cooperation from a positive or negative perspective in terms of the ability of a frame to achieve equity. This can be influenced by many factors and has the potential to change over time.

Influential factors on transboundary water cooperation: Scale and context

Transboundary water cooperation is place-based. Place is a way of understanding the world (Cresswell 2015), and this understanding of a place and the way we think about a place impact the formation and structure of cooperation, as well as what elements of cooperation are relevant for a particular place. While there are many aspects, such as social and cultural constructions, that affect the ideas of place and its relationship to cooperation, two have been identified from the interview and literature data that influence transboundary water cooperation in the context of the Four Frames of Cooperation: scale and context.

Scale

Scale is a social construction through political processes that establish a hierarchy of scales (Graddy 2011). Scale is the territorial delineation of a biological or ecological landscape or of a social, political, or economic entity, which incorporates the spatial dimension into transboundary water cooperation. In the case of transboundary water cooperation, scale can be considered as having two dimensions – the geographic scale and the governance scale. Shared waters, particularly river systems, are hierarchical and nested in nature, with smaller tributaries draining a landscape and converging into larger rivers flowing to a common outlet at an inland body of water, sea, or the ocean; see Figure 6.8

Figure 6.8 Columbia River Basin and Sub-basins.

This map shows the Columbia River Basin with on the sub-basins – the Willamette River. The Willamette River can be further broken down into its own sub-basins.

for an example of the nested sub-basins within the Columbia River Basin. The geographic extent of these nested basins and sub-basins corresponds to the geographic scope of transboundary water cooperation. Groundwater, given its complexities, does not neatly align with the idea of nested hierarchical scales, unlike surface water; however, geographic extent, recharge and discharge areas, and the potential for interconnectedness both vertically and horizontally in the subsurface affect how scale is perceived and applied to cooperation over shared groundwaters. The scale of governance embodies political geography's understanding and inclusion of power: the international system, supranational organizations, nation-states, regional governments, city, community, and local scales. Like the nested hierarchy of the geographic scales, governance scales are generally perceived based on physical size and power, where larger size is equated with greater power (Graddy 2011). This view can devalue cooperation occurring at smaller scales. Conventionally, we associate large geographic scales – like river basins – with *higher* governance scales, such as the nation-state or international organizations. However, when conceptualizing transboundary water cooperation using the Four Frames of Cooperation, the geographic and governance scales should not be considered as linked but rather evaluated with respect to cooperation in their own right. Furthermore, global politics and the rise of institutionalism and the global civil society are changing the thinking on conventional nested hierarchical scales. The powers that the states have traditionally maintained are shifting to international organizations, regional and community levels, along with non-state actors and civil society (Bulkeley 2005). This change is helping to decouple the geographic scale and the governance scale in transboundary water cooperation while ideally encouraging the inclusion of non-state actors into transboundary water cooperation processes.

Scale is a key aspect of international water law, more commonly identified as geographic scope (Su 2014). For example, Wouters (2013) uses scope as one of five measures for analyzing the operation and effectiveness of transboundary water management. She defines scope as determining the "geographical and hydrological parameters or limits of the treaty's application by defining both the water resources governed and the states eligible to participate" (Wouters 2013). Without defining the scope of an agreement, it is difficult to cooperate effectively (Wouters and Chen 2013). Often, the legal scope is expanded to include issue areas covered by the agreement. However, with respect to the Four Frames of Cooperation, I am using the conceptualization provided by Wouters (2013) in defining scope by the geographic and governance stakeholder limits. Therefore, I propose using the two types of scale presented in the above paragraph – geographic and governance – within the Four Frames of Cooperation, see Figure 6.9 for a representation of how these two scales can be sub-divided. How the geographic and governance scales are defined for a particular shared water will impact the format and function of the cooperative process. This will be apparent in what frames of cooperation the shared water has, as well as the types of key elements found within those frames. Let us consider a brief example where we are defining the cooperation between India and

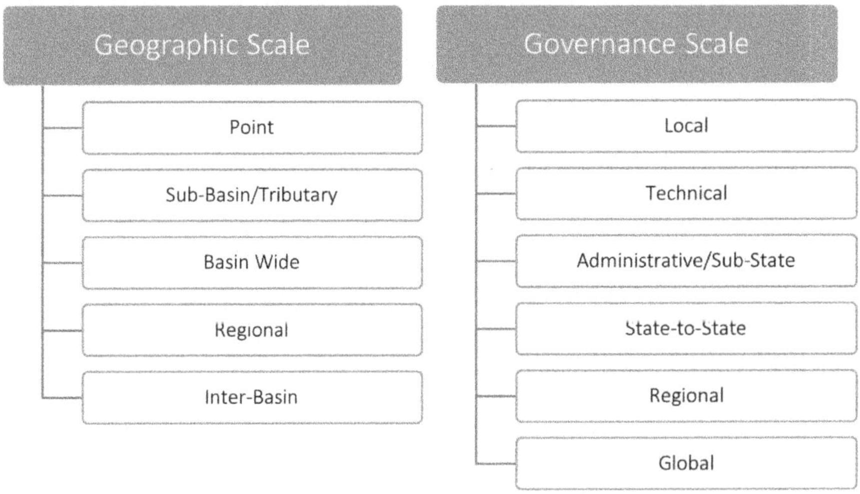

Figure 6.9 Geographic and governance scales.
These subdivisions create a common language for discussing scale; however, they should not restrict other conceptualizations of scale.

Bangladesh over the Ganges. This is occurring at a sub-basin-wide geographic scale and state-to-state governance scale. This compares to defining the cooperation between the twin cities of Nogales in Arizona, USA, and Sonora, Mexico, which share piped water across the border. Cooperation, in this case, is at the point geographic scale and the local governance scale. The frames that comprise transboundary water cooperation in each of these cases may differ due to the scale, as will the key elements that define each of the frames. Therefore, when using the Four Frames of Cooperation to define transboundary water cooperation, we must be aware of the geographic and governance scales, and how these have shaped the process of cooperation. This will affect our ability to interpret the definition and the level of transboundary water cooperation as it relates to scale.

Context

Similar to scale, which could arguably be included as an aspect of context, context also influences how the definition of transboundary water cooperation will differ in various shared waters. In addition, context affects how key elements are represented within the Four Frames of Cooperation. It may seem glaringly obvious to mention context as an influential factor on transboundary water cooperation; however, I believe it is particularly important given that context has the ability to shape and transform the structures and functions of cooperation. Contextual explanations of cooperation are difficult to systematize, as they cover a broad range, which is why they have only rarely been considered when accounting for cooperation (Lindemann 2008). Generally, power, interest, and knowledge-based arguments are presented in the study of cooperation (Lindemann 2008). While this book does not conduct a systematic evaluation of contextual factors influencing cooperation as Lindemann suggests, I do assert that context should be considered in the definition and evaluation of cooperation in the context of the Four Frames of Cooperation.

Most commonly, the context includes geography, hydrology, climate, and politics. Le Marquand (1977) identifies hydrological, economic, foreign policy, and domestic dimensions of shared water resources. Other authors have addressed context in their work – some examples: Conti (2014) includes previous water cooperation as an enabling factor for transboundary water cooperation; Kistin (2007, 2009) reminds readers that the historical context that goes beyond treaties and agreements needs to be included in regime formation and implementation; and Chokkakula (2018) identifies three enabling factors for cooperation – institutions, political relations, and the historical and geographical context. In terms of how these contextual factors influence the definition of cooperation through the four frames, each basin's unique context will help determine the appropriate key elements to define the four frames, or it may hinder the fulfillment of all four frames. For example, the upstream and downstream positioning of riparians in a basin has the potential to create incentives and disincentives for cooperation, altering the types of outcomes established between the parties or potentially creating a power imbalance resulting in low trust in the relational frame of cooperation. The *Indus River Treaty* provides a clear example of how the geography of the river influenced the division of tributaries between India and Pakistan. Some scholars believe this geographical division has led to increased resilience of the treaty and cooperation between India and Pakistan despite the tumultuous – often hostile – political relationship between the states (Chokkakula 2018; Iyer 2007). This brief example shows the intersection of geographic, historical, and political aspects of the context within transboundary water cooperation through the *Indus River Treaty* and the three frames it addresses – legal, relational, and outcome.

Within the most commonly included areas of context presented above, there are several specific contextual factors that appear in the literature and interview data and are worth discussing briefly here: meanings of water, variability, risk, and international donors and organizations.

First, one of the challenges at the core of cooperation is reconciling different meanings of water and its use. Because of the unique context each actor or stakeholder is part of in a cooperative process, there is the potential for differing and contested meanings of water (Aggestam 2015; Mollinga 2008). The competing understandings, positions, and interests have the potential to inhibit or promote cooperation, affect relationships, steer the legal and institutional mechanisms developed, and determine the outcomes. In the data, both the literature and interviewees stressed the importance of developing common understandings on a shared resource; this could range from a shared hydrologic or technical model of the resources (i.e., watershed model or conceptual model of an aquifer) to more abstract shared values and beliefs over water (i.e., the Four Worlds Framework for understanding needs, see Wolf 2010, 2017). Identifying common problems and understandings of a shared resource has significant potential to influence water diplomacy during negotiations and transboundary water cooperation. Shared meanings of water's influence on cooperation are captured by the relational frame of cooperation in the Four Frames of Cooperation.

The next contextual factor that is called out in the data is variability. Flow and supply variability, particularly due to climate change, is an influential factor in cooperation. Flow variability and supply variability fall within the categories of hydrological and climate context mentioned previously but are worth mentioning explicitly here, as they stood out in the data. Studies have shown that changes in flow can impact the stability of cooperative arrangements (Bhaduri et al. 2011), as well as drive cooperation to occur (Dinar et al. 2010, 2015). While higher variability in precipitation volume and timing is expected with climate change, this increased variability is likely to increase cooperation up until a threshold, after which it has the potential to encourage non-cooperative or unilateral behavior (Dinar et al. 2015; Intelligence Community Assessment 2012). We also know that the potential for conflict can increase if there is a rapid environmental change and not enough institutional capacity to absorb the change (Wolf, Yoffe, and Giordano 2003). Therefore, the risk of variability has the potential to encourage elements within the legal, institutional, and relational frames of cooperation, in the interest of preventing future conflict. Furthermore, the relational frame, I believe, will be particularly important as basin and aquifer management adapts from *conventional* command-and-control management that is reliant on predictability to more adaptive management. Trust, communication, and relationships will be vital aspects in the capacity for states and other stakeholders to address future variability. Like variability, risk also drives stakeholders' willingness to cooperate (and is mentioned in the data). Subramanian et al. (2012, 2014) have studied perceived risk and risk reduction and their impact on cooperation. Perceived risk is defined as the perception that cooperation could lead to a state's harm, including the loss of something of value or the political future of the state (Subramanian, Brown, and Wolf 2012). Therefore, to encourage participation in cooperation, states and other stakeholders must see a reduction of the perceived risk. The authors further define categories of risk (see Subramanian, Brown, and Wolf 2014) but, ultimately, these are heavily dependent on the context of the basin or aquifer. One example mentioned several times in the data, particularly in the interviews, is the risk of re-negotiation. The interviewees explained that states needed legal and institutional mechanisms to establish a formal structure of cooperation that aimed to prevent the need for re-negotiation under changing conditions. The Jordan River Basin was given as an example where the risk of re-negotiation factored into the development and negotiation of the *Treaty of Peace* in 1994 for the Jordan River between Jordan and Israel. The historical and political context in the basin has contributed to the low trust in the relational frame of cooperation. This contributes to a perceived risk of re-negotiation in the future, highlighting the need for clearly defined mechanisms allocating water between the countries.

Lastly, the final contextual factor that was called out in the data is the role and influence of international donors and organizations in promoting, developing, and maintaining transboundary water cooperation. International organizations and donors play an important role in helping to facilitate cooperation as third parties (Albrecht et al. 2018). In addition, regional frameworks also help to provide incentives for cooperation (Bakker 2009). International donors and financial institutions can provide funding or compensation to encourage unwilling participants into the process (Bernauer 2002). Furthermore, they can foster cooperation by funding and implementing transboundary water projects, infrastructure, and institutions (Blumstein 2017). In defining cooperation, particularly with reference to context, we need to consider how these third parties influence the key elements of cooperation that are developed within the Four Frames of Cooperation. Depending on the context, key elements encouraged and supported by donors and international agencies may be appropriate for the basin or aquifer, or they could be adapted to fit better the context and needs of the shared waters and its stakeholders, rather than that of the donors or international organizations. For example, Lautze and Giordano (2005) note that water cooperation in Africa is heavily influenced by external actors, which may have the tendency to diminish the organic development of cooperative processes. Similarly, Soliev et al. (2015) argue that the arrangements brokered by donors in the Aral Sea basin did not engage with the larger picture, leading to a simplification of issues and potentially a decline in cooperation. Several interviewees acknowledged that there can be a conflict between donor-driven and state-driven needs for a cooperative process, particularly when a project is being funded by third parties. Continued financing of cooperative processes or transboundary projects was also brought up as an influence on cooperation by interviewees, both as a positive aspect and as a negative aspect. Funding of full projects by third parties can help to break political barriers and encourage trust between stakeholders sharing a water body (Chen, Pernetta, and Duda 2013); however, several practitioners, in particular, noted that donor funding is not a reliable source for long term cooperative processes. Without adequate financial support from the shared water body's stakeholders, cooperative processes are often less effective or evaporate over time without continued third-party support. This discussion is not meant as a criticism of the influence of donors and international agencies but rather to encourage critical analysis when we define cooperation using the Four Frames of Cooperation in order to identify better the key elements established within the context of the basin or aquifer and those that may serve the needs of third parties. By understanding this distinction, we may be better able to understand the role of third parties in cooperation and encourage more fit-for-context elements of cooperation. The Global South has had significantly more external pressure and influence from third parties than the cooperative processes in the Global North, which have been typically spurred by internal forces (Rogers and Hall 2003); this could have had an impact on the types of key elements of cooperation established within the Four Frames. Therefore, further studies considering the intersection of third parties – such as donors and international organizations – with context and the formation and effectiveness of cooperation are needed.

Application of the Four Frames of Cooperation: Two examples

Now that we have discussed the frames and the main structures of the Four Frames of Cooperation framework, how would one apply the framework to develop a definition of a transboundary water cooperative process? When applying the four frames to a specific place, several steps can be used. These are depicted as a flowchart in Figure 6.10, which summarizes the full Four Frames of Transboundary Water Cooperation, which provides examples of key elements in each of the four frames. These examples demonstrate how key elements of cooperation vary with scale or context. Prior to applying the framework, extensive research and understanding of the basin or aquifer are needed. This can include desk-based research and, ideally, interviews and on-the-ground data collection. Interviews and in basin data collection will particularly aid in identifying key elements within the relational frame. The general steps are as follows:

1. To begin, identify the cooperative process of interest. For example, the process of interest could be the transboundary water cooperation occurring between states over the operation of a dam, allocation of surface water, or water quality protections of an aquifer.
2. Once you have identified your process of interest, the second step is to identify the scale or scales at which the cooperation is occurring. For example, if one were to define the cooperation between the United States and Mexico over the Colorado River, the scale of the process of interest could be at the basin-wide level and address the state-to-state governance. An alternative example is if one were to define the cooperation over the operation of a dam, such as the Arpacay/Akhuryan Dam on the river border between Turkey and Armenia. The scale of the process of interest could be at the point scale and address the state-to-state and technical governance levels. By identifying the scale(s), we can set boundaries to help determine what is or is not relevant to the cooperative process we are defining. This includes identifying both the geographic and governance scales, noting that while often these are paired, there is potential for multi-scalar processes as well as uncoupled hierarchies between the geographic and governance scales.
3. The next step is to explore the context in which the cooperative process of interest occurs at the scales identified. What is known about the geography,

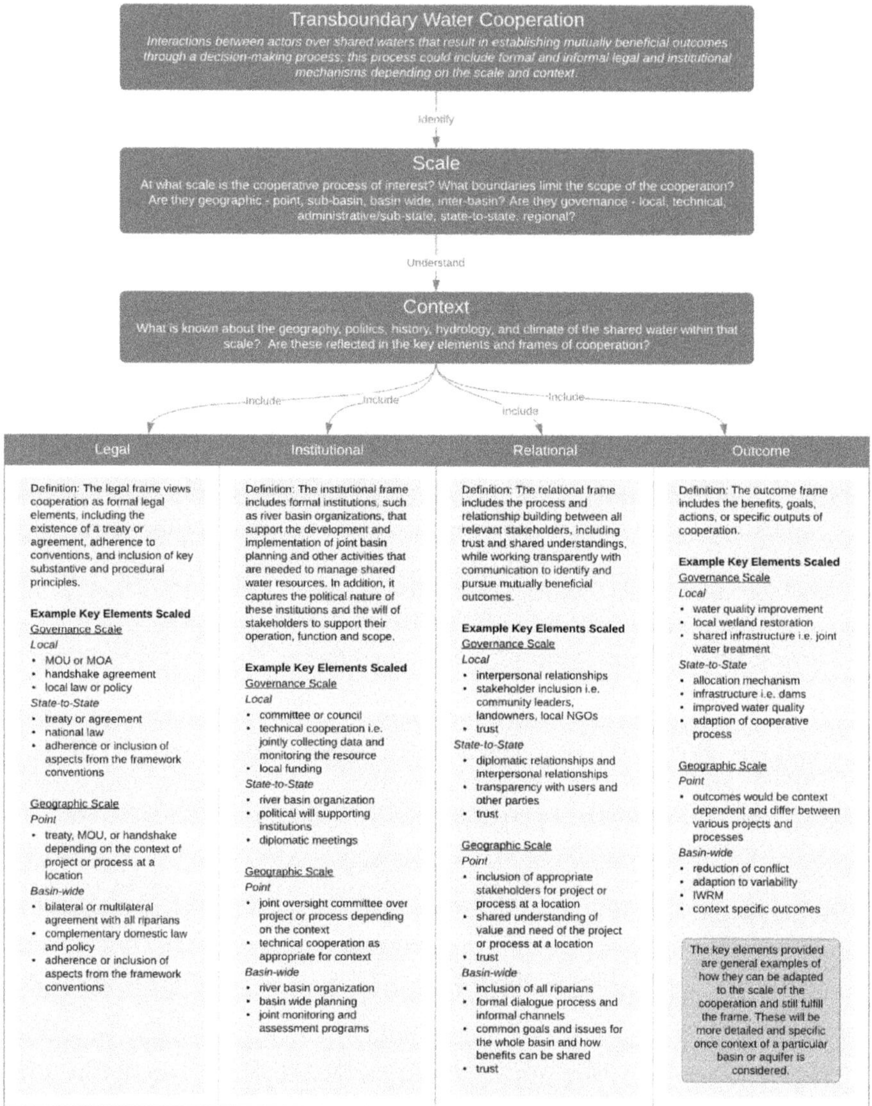

Figure 6.10 The Four Frames of Transboundary Water Cooperation.

This graphic presents the framework, factors for consideration that will influence cooperation, and examples of how scale will alter what key elements are within each frame. This framework is neutral and is developed to be used as a tool to define cooperation in an explanatory manner in a particular basin, aquifer, and so on.

hydrology, climate, political situation, or history? In particular, consider the meanings of water, variability and risk, and international donors and organizations involved or related to the cooperative process. As discussed in the above section, these factors are prominent contextual factors influencing cooperation from the data that might not initially be considered when determining the contextual factors.

4. Once these boundaries (process of interest, scale) have been established, the key elements within each of the four frames can be identified using the extensive research on the basin or aquifer through the contextual lens of the basin. Using the key elements as a structure, the scope and extent of the cooperative process within each frame can be explained. By structuring a context-specific definition of transboundary water cooperation in this way, we can identify areas within the frames where cooperation is or is not happening. In addition, we can determine if the cooperative process is at the appropriate scale, what might be missing, and if the identified key elements align with the context.

While following this process, it is essential to remember that this framework is neutral. Therefore, the next step is to evaluate and measure cooperation, which is the focus of the following Part III. However, before diving into the evaluation of cooperation's effectiveness, I want to illustrate the framework through two brief examples – Lake Victoria and the Goascorán River Basin. These examples draw from the background provided in Chapter 3, where the extent of the transboundary area with an operational arrangement for cooperation was calculated for Uganda and Honduras. What is important to note about these examples is that the framework can be applied in substantially different contexts but still capture the key elements and balance between the four frames due to its flexibility and ability to adapt to each place.

Defining cooperation in the Lake Victoria Sub-Basin

The Lake Victoria Basin includes Uganda, Rwanda, Burundi, Tanzania, and Kenya; the lake resides within Uganda, Tanzania, and Kenya. In 1999, the East African Community (EAC) was established between the basin states. Under the regional agreement, the three lake area states signed *the Protocol for Sustainable Development of Lake Victoria Basin* in 2003, which created the Lake Victoria Basin Commission. Considering the transboundary water cooperation on Lake Victoria, we can clearly see that this cooperative process has both legal and institutional components. The legal frame is addressed by the regional EAC agreement and the protocol, and the institutional frame includes the EAC Community and the basin commission. If a deeper analysis of the cooperation at Lake Victoria was within the scope of this work, we could identify the specific legal and institutional mechanisms that further provide support for the inclusion of the two frames in the definition of transboundary water cooperation in the basin.

Outcomes have been identified more generally by the protocol for sustainable use and management of the basin and more specifically in the joint projects developed through the basin commission, such as the Lake Victoria Environment Management Programme (LVEMP) I and II. The LVEMP II set joint objectives to improve the management of the lake and addresses pollution hotspots for the benefit of the communities that depend on the lake (Lake Victoria Basin Commission 2016). The joint project's objectives highlight one example of outcomes determined by the riparian members of the basin commission and donors. The relational aspect of the cooperation in the basin is difficult to determine here given the desk-based survey of cooperation presented in Chapter 3; however, we can speculate that there are shared understandings and common goals identified between the stakeholders, given the types of outcomes developed in the LVEMP. This example is not meant to do a deep dive into the transboundary water cooperation within the Lake Victoria Basin. However, it aims to illustrate how the framework can be used to identify the various frames within a cooperative process and potentially identify elements that could further transboundary water cooperation. With this framework, I am not aiming to pass a value judgment on a cooperative process but merely to provide a definition for clarification and consistent conceptualization of transboundary water cooperation.

Defining cooperation in the Goascorán River Basin

The Goascorán River Basin is shared between Honduras and El Salvador. The Goascorán drains to the Pacific Ocean and forms part of the border between the two states. It differs from Lake Victoria in that there is no formal treaty or agreement governing cooperation between the states. However, third-party donors have been heavily involved in the basin in recent years, mainly through the IUCN BRIDGE Program. For additional background on the Goascorán, see Chapter 3. For this example, we consider the basin-wide transboundary management and cooperation occurring in the basin at the basin-wide geographic scale and in the state-to-state, sub-state, and technical governance scales.

Within the legal frame, the Goascorán differs significantly from the Lake Victoria sub-basin. There is neither a regional nor interstate agreement that directly covers the basin. However, Honduras does have a national water law established in 2009 that incorporated IWRM principles (GWP Central America 2015). In the basin specifically and driven by the local communities, border development agencies and projects were developed beginning in 2004 financed by the European Union; this included governance at both the national and local scales. Within the scope of this development work, the Bi-National Management Group of the Goascorán River Basin was established in 2007 (Maier et al. 2016). Therefore, while the legal elements of cooperation are not present in the basin, there is still an established institutional mechanism. Initially, this Bi-National Management Group of the Goascorán River Basin only included local governments and civil society and was suspended in 2009 due to the political situation in Honduras. The group was resurrected in 2011 through support from IUCN's BRIDGE Project and their partner organizations (Maier et al. 2016; MacQuarrie et al. 2013). Within the institutional frame, we can see that the influence of donors and donor funding has impacted the continuity of cooperation in the basin. While local government and civil society are willing to cooperate and jointly manage the basin, the lack of national government support and funding has potentially limited the group's capacity, leading to the reliance on third-party donors. This example illustrates the influence of scale and context on the institutional frame of cooperation.

Concerning the outcome frame, a Basin Management Plan was developed in 2007. However, this plan was limited in execution, and as noted by the BRIDGE project, maybe due to the lack of a managing entity (Maier et al. 2016). While joint management objectives have been established through the plan, the outcome frame has limited key elements at all scales and is influenced by the extent of cooperation in the legal and institutional frames. Based on the work of the BRIDGE project, it is noted that there are different levels of trust depending on the actors. For example, there is moderate trust in municipalities and local government bodies, yet low trust in public sector actors. However, it was also noted that not all the relevant actors are involved in the Bi-National Management Group of the Goascorán River Basin (Maier et al. 2016; MacQuarrie et al. 2013). From this brief look at the trust and actor involvement in the cooperative process, we can better understand the relational elements of cooperation and how these elements link with the institutional and outcome frames. Like with the Lake Victoria example, this is a brief discussion of how the Four Frames of Transboundary Water Cooperation can provide a structure for defining cooperation and a mechanism for exploring how different elements are interrelated, vary with scale, and influenced by context, without making a value judgment. Further, these two examples demonstrate the flexibility and recognition of place that the framework allows.

Conclusion

This chapter presents the Four Frames of Cooperation as a framework for defining cooperation with respect to a particular place, scale, and context. This framework defines transboundary water cooperation as *interactions between actors over shared waters that result in establishing mutually beneficial outcomes through a decision-making process; this process could include formal and informal legal and institutional mechanisms depending on the scale and context.* In order to make this broad and general definition more easily applicable and usable for defining cooperation, it separates cooperation into four frames – legal, institutional, relational, and outcome. All four frames are represented in the above definition of transboundary water cooperation; however, not all cooperative processes may have all four frames, as demonstrated by the Goascorán example. I argue that for full transboundary water cooperation to exist, all Four Frames of Cooperation should be represented at the appropriate scale for the context of the shared basin or aquifer. Cooperation in this framework is neutral. It does not aim to pass a value judgment or evaluate the cooperation occurring – this is the goal of Part III. It merely attempts to create a straightforward method for consistently defining cooperation while allowing for differences in scale and context. Part III defines effective cooperation and presents a model that builds on the Four Frames of Cooperation to evaluate the effectiveness of a cooperative process.

Notes

1 Please keep in mind when reading that this chapter focuses on defining *transboundary water cooperation*, not cooperation in the broader sense of the word. I use *cooperation* without the *transboundary water* modifier; however, I am referring to *transboundary water cooperation*. I will note when I refer to cooperation unlinked to transboundary water.
2 Expert interviewees included 22 Academics, 23 Practitioners, 3 Academic/Practitioners, and 6 Practitioners and Academics affiliated with NGOs: equaling a total of 54 interviews from 27 countries.

References

Abdullaev, Iskandar. 2018. Research Interview.
Aggestam, Karin. 2015. "Desecuritisation of Water and the Technocratic Turn in Peacebuilding." *International Environmental Agreements: Politics, Law and Economics* 15 (3): 327–40. https://doi.org/10.1007/s10784-015-9281-x.
Albrecht, Tamee R., Robert G. Varady, Adriana A. Zuniga-Teran, Andrea K. Gerlak, Rafael Routson De Grenade, América Lutz-Ley, Facundo Martín, et al. 2018. "Unraveling Transboundary Water Security in the Arid Americas." *Water International* 43 (8): 1075–1113. https://doi.org/10.1080/02508060.2018.1541583.
Bakker, Marloes H.N. 2009. "Transboundary River Floods and Institutional Capacity." *JAWRA Journal of the American Water Resources Association* 45 (3): 553–66. https://doi.org/10.1111/j.1752-1688.2009.00325.x.
Bernauer, Thomas. 2002. "Explaining Success and Failure in International River Management." *Aquatic Sciences* 64 (1): 1–19. https://doi.org/10.1007/s00027-002-8050-4.
Bernauer, Thomas, and Tobias Siegfried. 2008. "Compliance and Performance in International Water Agreements: The Case of the Naryn/Syr Darya Basin." *Global Governance* 14 (4): 479–501.
Bhaduri, Anik, Utpal Manna, Edward Barbier, and Jens Liebe. 2011. "Climate Change and Cooperation in Transboundary Water Sharing: An Application of Stochastic Stackelberg Differential Games in Volta River Basin." *Natural Resource Modeling* 24 (4): 409–44. https://doi.org/10.1111/j.1939-7445.2011.00097.x.
Blumstein, Sabine. 2017. "Managing Adaptation: International Donors' Influence on International River Basin Organizations in Southern Africa." *International Journal of River Basin Management* 15 (4): 461–73. https://doi.org/10.1080/15715124.2017.1339354.
Blumstein, Sabine, and Susanne Schmeier. 2017. "Disputes over International Watercourses: Can River Basin Organizations Make a Difference?" In *Management*

of Transboundary Water Resources under Scarcity, edited by Ariel Dinar and Yacov Tsur, 191–236. New Jersey: World Scientific Publishing Company.

Brochmann, Marit. 2012. "Signing River Treaties—Does It Improve River Cooperation?" *International Interactions* 38 (2): 141–63. https://doi.org/10.1080/03050629.2012.657575.

Bulkeley, Harriet. 2005. "Reconfiguring Environmental Governance: Towards a Politics of Scales and Networks." *Political Geography* 24 (8): 875–902. https://doi.org/10.1016/j.polgeo.2005.07.002.

Caponera, Dante A. 1985. "Patterns of Cooperation in International Water Law: Principles and Institutions." *Natural Resources Journal* 25: 26.

Chen, Sulan, John C. Pernetta, and Alfred M. Duda. 2013. "Towards a New Paradigm for Transboundary Water Governance: Implementing Regional Frameworks through Local Actions." *Ocean & Coastal Management* 85 (December): 244–56. https://doi.org/10.1016/j.ocecoaman.2012.10.019.

Chokkakula, Srinivas. 2018. "Transboundary Politics of Cooperation: Telugu Ganga Project, India." *Regional Environmental Change* 18 (6): 1645–54. https://doi.org/10.1007/s10113-018-1348-0.

Conti, Kirstin. 2014. "Factors Enabling Transboundary Aquifer Cooperation: A Global Analysis." Deft, The Netherlands: IGRAC. http://citeseerx.ist.psu.edu/viewdoc/download?doi=10.1.1.961.1937&rep=rep1&type=pdf.

Cresswell, Tim. 2015. *Place: An Introduction*. Second edition. Chichester, West Sussex, UK; Malden, MA: Wiley & Sons. http://search.ebscohost.com/login.aspx?direct=true&scope=site&db=nlebk&db=nlabk&AN=842630.

De Stefano, Lucia, James Duncan, Shlomi Dinar, Kerstin Stahl, Kenneth Strzepek, and Aaron Wolf. 2010. "Mapping the Resilience of International River Basins to Future Climate Change-Induced Water Variability." Water Sector Board Discussion Paper Series 15. Washington, D.C.: The World Bank.

De Stefano, Lucia, Jacob D. Petersen-Perlman, Eric A. Sproles, Jim Eynard, and Aaron T. Wolf. 2017. "Assessment of Transboundary River Basins for Potential Hydro-Political Tensions." *Global Environmental Change* 45 (July): 35–46. https://doi.org/10.1016/j.gloenvcha.2017.04.008.

Devlaeminck, David. 2018. "Revisiting the Substantive Rules of the Law of International Watercourses: An Analysis through the Lens of Reciprocity and the Interests of China." *Water Policy* 20 (2): 323–35. https://doi.org/10.2166/wp.2017.069.

Dinar, Ariel, Brian Blankespoor, Shlomi Dinar, and Pradeep Kurukulasuriya. 2010. "Does Precipitation and Runoff Variability Affect Treaty Cooperation between States Sharing International Bilateral Rivers?" *Ecological Economics* 69 (12): 2568–81. https://doi.org/10.1016/j.ecolecon.2010.07.036.

Dinar, Shlomi, David Katz, Lucia De Stefano, and Brian Blankespoor. 2015. "Climate Change, Conflict, and Cooperation: Global Analysis of the Effectiveness of International River Treaties in Addressing Water Variability." *Political Geography* 45 (March): 55–66. https://doi.org/10.1016/j.polgeo.2014.08.003.

Earle, Anton, and Patricia Wouters. 2015. "Implementing Transboundary Water Cooperation through Effective Institutional Mechanisms." *The Journal of Water Law* 24: 100–114.

Entman, Robert. 1993. "Framing: Toward Clarification of a Fractured Paradigm." *Journal of Communication* 43 (4): 51–58.

Gerlak, Andrea K. 2018. Research Interview.

Goffman, E. 1974. *Frame Analysis*. Cambridge: Harvard University Press.

Graddy, Garrett. 2011. "Politics of Scale." In *Green Politics: An A-to-Z Guide*, edited by Dustin Mulvaney, 328–31. Los Angeles: Sage Publications.

GWP Central America. 2015. "Situacion de las Recursos Hidricos en Centroamerica: Honduras." Tegucigalpa, Honduras: Global Water Partnership.

Herbert, Anne L. 1996. "Cooperation in International Relations: A Comparison of Keohane, Haas and Franck." *Berkeley Journal of International Law*. https://doi.org/10.15779/z38135k.

Ide, Tobias, and Adrien Detges. 2018. "International Water Cooperation and Environmental Peacemaking." *Global Environmental Politics* 18 (4): 63–84. https://doi.org/10.1162/glep_a_00478.

Intelligence Community Assessment. 2012. "Global Water Security." Washington, D.C.: Office of the Director of National Intelligence.

Iyer, Ramaswamy R. 2007. *Towards Water Wisdom: Limits, Justice, Harmony*. New Delhi: SAGE Publications.

Keohane, Robert O. 1984. *After Hegemony: Cooperation and Discord in the World Political Economy*. Princeton: Princeton University Press. https://academic.oup.com/ia/article-lookup/doi/10.2307/2617490.

Kistin, Elizabeth J. 2007. "Trans-Boundary Cooperation in SADC: From Concept to Implementation." In *8th WaterNet/WARFSA/GWP-SA Symposium,* Lusaka, Zambia. Vol. 30.

Kistin, Elizabeth J., Peter J. Ashton, Anton Earle, Daniel Malzbender, Marian J. Patrick, and Anthony R. Turton. 2009. "An Overview of the Content and Historical Context of the International Freshwater Agreements That South Africa Has Entered into with Neighbouring Countries." *International Environmental Agreements: Politics, Law and Economics* 9 (1): 1–21. https://doi.org/10.1007/s10784-008-9086-2.

Lake Victoria Basin Commission. 2016. "LVEMP II | LVBC." East African Community Lake Victoria Basin Commission. October 29, 2016. www.lvbcom.org/node/48.

Lautze, Jonathan, and Mark Giordano. 2005. "Transboundary Water Law in Africa: Development, Nature, and Geography." *Natural Resources Journal* 45: 36.

LeMarquand, David G. 1977. *International Rivers: The Politics of Cooperation*. Vancouver: Westwater Research Centre, University of British Colombia.

Lindemann, Stefan. 2008. "Understanding Water Regime Formation—A Research Framework with Lessons from Europe." *Global Environmental Politics* 8 (4): 117–40. https://doi.org/10.1162/glep.2008.8.4.117.

MacQuarrie, Patrick, Rebecca Welling, Nazareth Porras, and Rocio Córdoba. 2013. "Goascorán Basin: Honduras and El Salvador." Gland, Switzerland: IUCN.

Maier, Luis, Nazareth Porras, Roco Cordoba, Patrick MacQuarrie, and Rebecca Welling. 2016. "The Goascoran River Basin: Honduras and El Salvador." BRIDGE Case Study. San José, Costa Rica: International Union for Conservation of Nature. www.iucn.org/sites/dev/files/goascoran_-_english.pdf.

Mianabadi, Hojjat, Erik Mostert, Saket Pande, and Nick van de Giesen. 2015. "Weighted Bankruptcy Rules and Transboundary Water Resources Allocation." *Water Resources Management* 29 (7): 2303–21. https://doi.org/10.1007/s11269-015-0942-x.

Mirumachi, Naho. 2007. "Introducing Transboundary Waters Interaction Nexus (TWINS): Model of Interaction Dynamics in Transboundary Waters." In *Third International Workshop on Hydro-Hegemony*, 12–13. www.lmei.soas.ac.uk/water/publications/papers/file39696.pdf.

Mirumachi, Naho, and John Anthony Allan. 2007. "Revisiting Transboundary Water Governance: Power, Conflict, Cooperation and the Political Economy." In

International Conference on Adaptive and Integrated Water Management, 12–15. Citeseer. http://citeseerx.ist.psu.edu/viewdoc/download?doi=10.1.1.490.9078&rep=rep1&type=pdf.

Mollinga, Peter P. 2008. "Water, Politics and Development: Framing a Political Sociology of Water Resources Management" *Water Alternatives* 1 (1): 7–23.

Molnar, Kata, Rosa Cuppari, Susanne Schmeier, and Siegfried Demuth. 2017. "Preventing Conflicts, Fostering Cooperation – The Many Roles of Water Diplomacy." Stockholm: UNESCO's International Centre for Water Cooperation.

Mthimkhulu, Sindy. 2017. Research Interview.

Ostrom, Elinor. 1990. *Governing the Commons*. Cambridge: Cambridge University Press.

———. 2005. *Understanding Institutional Diversity*. Princeton: Princeton University Press. https://press.princeton.edu/titles/8085.html.

———. 2008. "Institutions and the Environment." *Economic Affairs* 28 (3): 24–31. https://doi.org/10.1111/j.1468-0270.2008.00840.x.

———. 2010. "Institutional Analysis and Development: Elements of the Framework in Historical Perspective." In *Historical Developments and Theoretical Approaches in Sociology - Volume II*, edited by Charles Crothers, 261–88. Paris, France: UNESCO-EOLSS Publications.

Rogers, Peter, and Alan W Hall. 2003. *Effective Water Governance*. Technical Committee Background Papers 7. Stockholm: Global Water Partnership.

Sallisalmi, Virve. 2018. Research Interview.

Schmeier, Susanne. 2013. *Governing International Watercourses: River Basin Organizations and the Sustainable Governance of Internationally Shared Rivers and Lakes*. New York: Routledge. www.routledge.com/Governing-International-Watercourses-River-Basin-Organizations-and-the/Schmeier/p/book/9780415623582.

———. 2015. "The Institutional Design of River Basin Organisations: Empirical Findings from around the World." *International Journal of River Basin Management* 13 (1): 51–72.

Sindico, Francesco. 2018. Research Interview.

Soliev, Ilkhom, Kai Wegerich, and Jusipbek Kazbekov. 2015. "The Costs of Benefit Sharing: Historical and Institutional Analysis of Shared Water Development in the Ferghana Valley, the Syr Darya Basin." *Water* 7 (6): 2728–52. https://doi.org/10.3390/w7062728.

Su, Yu. 2014. "Contemporary Legal Analysis of China's Transboundary Water Regimes: International Law in Practice." *Water International* 39 (5): 705–24. https://doi.org/10.1080/02508060.2014.950856.

Subramanian, Ashok, Bridget Brown, and Aaron T. Wolf. 2012. "REACHING ACROSS THE WATERS: Facing the Risks of Cooperation in International Waters." Water Papers. The World Bank. http://water.worldbank.org/sites/water.worldbank.org/files/publication/WaterWB-Reaching-Across-Waters.pdf.

———. 2014. "Understanding and Overcoming Risks to Cooperation along Transboundary Rivers." *Water Policy* 16 (5): 824. https://doi.org/10.2166/wp.2014.010.

TFDD. 2021. "Transboundary Freshwater Dispute Database." College of Earth, Ocean and Atmospheric Sciences, Oregon State University. 2021. https://transboundarywaters.science.oregonstate.edu/.

The Economist Intelligence Unit. 2020. "The Blue Peace Index 2020." London: The Economist. https://bluepeaceindex.eiu.com/pdf/EIU_Blue%20Peace%20Index_2020.pdf.

Tuomela, Raimo. 1993. "What Is Cooperation?" *Erkenntnis* 38 (1): 87–101. https://doi.org/10.1007/BF01129023.

Vollan, Björn, and Elinor Ostrom. 2010. "Cooperation and the Commons." *Science* 330 (6006): 923–24.

Wolf, Aaron T. 2010. *Sharing Water, Sharing Benefits: Working Towards Effective Transboundary Water Resources Management*. France: UNESCO-IHP.

———. 2017. *The Spirit of Dialogue: Lessons from Faith Traditions in Transforming Conflict*. Washington, DC: Island Press.

Wolf, Aaron T., Shira B. Yoffe, and Mark Giordano. 2003. "International Waters: Identifying Basins at Risk." *Water Policy* 5 (1): 29–60. https://doi.org/10.2166/wp.2003.0002.

Wouters, Patricia. 2013. "International Law–Facilitating Transboundary Water Cooperation." *Published by the Global Water Partnership, TEC Background Papers*, no. 17. http://papers.ssrn.com/sol3/papers.cfm?abstract_id=2363809.

Wouters, Patricia, and Huiping Chen. 2013. "China's 'Soft-Path' to Transboundary Water Cooperation Examined in the Light of Two UN Global Water Conventions - Exploring the 'Chinese Way.'" *The Journal of Water Law* 22: 19.

Zartman, I. William, ed. 2008. "Concept: Cooperation." *PIN Points* 30: 5–7.

Zeitoun, Mark, Ana Elisa Cascão, Jeroen Warner, Naho Mirumachi, Nathanial Matthews, Filippo Menga, and Rebecca Farnum. 2017. "Transboundary Water Interaction III: Contest and Compliance." *International Environmental Agreements: Politics, Law and Economics* 17 (2): 271–94. https://doi.org/10.1007/s10784-016-9325-x.

Zeitoun, Mark, and Naho Mirumachi. 2008. "Transboundary Water Interaction I: Reconsidering Conflict and Cooperation." *International Environmental Agreements: Politics, Law and Economics* 8 (4): 297–316. https://doi.org/10.1007/s10784-008-9083-5.

Zeitoun, Mark, Naho Mirumachi, and Jeroen Warner. 2011. "Transboundary Water Interaction II: The Influence of 'Soft' Power." *International Environmental Agreements: Politics, Law and Economics* 11 (2): 159–78. https://doi.org/10.1007/s10784-010-9134-6.

———. 2020. *Understanding Water Conflicts: Analysis for Transformation*. New York: Oxford University Press.

Part III

Challenge 2

What makes transboundary water cooperation effective?

7 Murky waters

Trying to define effectiveness for transboundary water cooperation

Part II defined transboundary water cooperation. It developed the Four Frames of Transboundary Water Cooperation, which is a neutral framework for developing an explicit definition of transboundary water cooperation in a particular context at a particular scale. Furthermore, Chapters 5 and 6 develop a generalized definition of transboundary water cooperation: *interactions between actors over shared waters that result in establishing mutually beneficial outcomes through a decision-making process; this process could include formal and informal legal and institutional mechanisms depending on the scale and context.* However, as discussed, any cooperation[1] is often identified as 'good.' Any cooperation is 'good.' But what does good mean? This assumption tends to ignore the actual outcomes and effects of cooperation (and conflict). Cooperation does not always establish the benefits it is expected to and, in fact, it can contribute to inequality and unsustainable use of the shared resource (Kistin 2007). Therefore, we need to begin evaluating cooperation to determine if it meets expectations, leads to improved management, reduced conflict, and so on (Daoudy and Kistin 2008). Ultimately, is the cooperation effective?

When we initially think of effective, the understanding that comes to mind is that there is a visible, measurable impact. In the literature, effectiveness is often seen as the ability to solve the problems that initiated the actors to enter into a cooperative process or establish a regime. However, when we look closer at what this actually means, we see that "effectiveness emerges as an elusive concept. It can mean a number of different things and some of its meanings require normative, scientific, and historical judgments" (Young and Levy 1999). Determining what effectiveness means in the context of transboundary water cooperation, therefore, requires careful examination. Through Chapters 7, 8, and 9, Part III answers this question: What makes transboundary water cooperation effective? Part III follows the same structure as Part II and approaches this question from two directions. This chapter first addresses how effective cooperation is defined in the literature, while Chapter 8 considers this question and develops a definition of effective transboundary water cooperation based on a qualitative content analysis of the literature and interview data. Chapter 9 brings the discussion to a method for evaluating effectiveness. By identifying how effective cooperation is perceived in the literature and within the sample

DOI: 10.4324/9781003126249-10

of expert interviewees, we can begin to develop a definition of effectiveness that can be used to distinguish better if cooperation is, in fact, 'good' by determining if it is effective.

Effectiveness in the broader literature

International relations literature has contributed significantly to the body of literature that conceptualizes and defines cooperation. Similarly, much of the literature on effectiveness stems from international relations, particularly from studies on international environmental politics, which is concerned with the effects of cooperation (O'Neill, Balsiger, and VanDeveer 2004). Regime theory has emerged as a method to explain cooperation in the anarchic international system (Daoudy and Kistin 2008; Keohane and Martin 1995). Regimes are understood to be a "set of transboundary level principles, norms, rules, and decision-making procedures around which actors' expectations converge in a given issue area" (Krasner 1983; Kranz, Menniken, and Hinkel 2010). Most commonly, regimes are discussed in context with international agreements and formal institutions; however, regimes can also exist in the absence of agreements (Levy, Young, and Zürn 1995). Under regime theory, institutions are generally seen as the tools to manage cooperation over natural resources; this differs from governance, as governance focuses more strongly on agent-based approaches to collective problems. These agents are often state actors in the case of international environmental problems (Phillips et al. 2006). Scholars have found that international regime theory presents a good starting point for explaining transboundary water cooperation. For example, Lindemann (2008) defines international water regimes as "sets of implicit or explicit principles, norms, rules, and decision-making procedures around which actor's expectations converge in the management of transboundary rivers." Much of the literature in international relations – particularly environmental politics – has historically focused on regime formation. There are two primary models of regime formation – utilitarian and power-based models. Utilitarian models of regime formation focus on states as rational actors who will work to reach an agreement for feasible joint gains if they are within a distinct zone of agreement. Institutions constructed, in this view, are unproductive, and states will work to realize the mutual benefits while keeping transaction costs low (Young 1989). On the other hand, the power-based model is based on realist perspectives, where cooperation is unlikely unless there is a hegemon. Institutional arrangements reflect the distribution of power in the system and, therefore, in the establishment of potential solutions for collective outcomes. The hegemon, according to realists, is critical for the establishment of an international regime or institution at the international level according to hegemonic stability theory (Young 1989; Lindemann 2008). Other authors have argued for additional theories for regime formation, such as through institutional bargaining (Young 1989). Lindemann (2008) proposes a framework for international water regime formation through drivers of power, interest, knowledge, and context. International regimes, ultimately, are formed as a result of international problems, which are context dependent and influenced by human actions. Both the context and the problem will affect the possible solutions, including the regime formed (Young 1999). Therefore, research questions have evolved from regime formation to assess how and why regimes succeed or fail – in other words, their effectiveness.

Despite the considerable efforts on effectiveness research, there is not a comprehensive model for defining and measuring the effectiveness of international regimes (Schmeier 2013). There are substantial differences in perspective on whether a regime or institution is successful and whether a governance system is effective (Young 2011). When applied to international regimes, the degree of complexity makes defining and measuring effectiveness a challenging task, and therefore what makes it effective can vary dramatically (Young 2011; Young and Levy 1999). Generally, there are two overlapping approaches to research on effectiveness. First, scholars attempt to determine what effectiveness means, along with methods for measuring the effectiveness of institutions (Helm and Sprinz 2000; Underdal 2002). The other area of research focuses on identifying and explaining the variables that establish the extent to which a regime is considered effective (Underdal 2002; M. Levy, Keohane, and Haas 1993). Hence, definitions of effectiveness vary. Broadly, effectiveness is understood as the *success* of the regime or institution (Underdal 1992; 2002; Bernauer 1995, 2002; O'Neill, Balsiger, and VanDeveer 2004; Young 2011; Helm and Sprinz 2000). This presents an issue, as what is meant by *success* will vary from regime to regime and with context. Therefore, effectiveness research is striving to define and establish a valid indicator of *success*. Previously, regime effectiveness was measured by the level of cooperation, which is defined in terms of the mechanisms implemented (i.e., enforcement and data exchange) (Underdal 1992; Kay and Jacobson 1983, 14–18). This method of identifying the level of cooperation is also prevalent in international waters literature; however, it is often strictly evaluating whether a mechanism is present in an arrangement regardless of its implementation. Similarly, others have been focusing on procedural indicators (Kay and Jacobson 1983, 316–17; Underdal 1992). This trends with the method developed to measure arrangements with operational cooperation in SDG Indicator 6.5.2. However, these methods are problematic, as we cannot link functions of arrangements or procedural indicators to *success* or effectiveness, particularly without a clear understanding of what we mean by effectiveness. The following Table 7.1 presents a few definitions of effectiveness from the international relations and regime effectiveness literature.

As seen in the definitions in Table 7.1, effectiveness is an elusive concept (Young and Levy 1999). Many of these definitions are vague or perhaps elegant in their simplicity but difficult to use as an analytical tool (Underdal 2002), particularly when different historical, political, and environmental aspects are considered, changing their meanings. Young and Levy (1999) attempt to capture this variability by defining several different approaches to effectiveness: problem solving, legal, economic, normative, and political. The

Table 7.1 Example of definitions of effectiveness or effective regimes in international relations and regime effectiveness literature

Definition	*Source*
Regime effectiveness is "the extent to which regimes contribute to solving or mitigating the problems that motivate those people who create the regime."	(Young 2011, 19854)
"Effectiveness is a matter of the contributions that institutions make to solving the problems that motivate actors to invest the time and energy needed to create them."	(Young and Levy 1999, 3)
"In a common-sense understanding, a regime can be considered effective to the extent that it successfully performs a certain (set of) function(s) or solves the problem(s) that motivated its establishment."	(Underdal 2002, 4)
"Successful institutions, in a very comprehensive sense of success, are those that (1) change the behavior of states and other actors in the direction intended by the cooperating parties, (2) solve the environmental problems they are supposed to solve, and (3) do so in an efficient and equitable manner."	(Bernauer 1995, 358)
"[e]ffectiveness is a function of structural constraints and opportunities and of the strategic choices of the actors involved."	(O'Neill, Balsiger, and VanDeveer 2004, 163)

problem-solving approach reflects much of the literature on effectiveness in that it argues that regimes develop as a result of a particular problem. Therefore, intuitively, effective regimes eliminate or alleviate the problem, but this definition is difficult to apply meaningfully (Young and Levy 1999). The legal definition of effectiveness is measured based on the degree to which legal obligations are adhered, including policies developed or implemented and rules complied with (Young and Levy 1999). Like the problem-solving approach, this presents challenges, as legal compliance does not equate to positive impacts. The economic approach incorporates the legal definition with efficiency – does the regime have compliance within the lowest transaction costs (Young and Levy 1999). Again, this approach has the same limitations as the legal definition, as the most economically efficient regime does not necessarily result in positive impacts. The normative approach is probably the least considered by scholars, as it is difficult to measure. This definition is based on normative principles, such as "fairness, justice, stewardship, participation, etc." (Young and Levy 1999). Lastly, the political approach to effectiveness considers international problems as a result of the "constellation of actors, interests, and institutions or behavioral complexes" (Young and Levy 1999). Therefore, a regime is considered effective if actors change their behaviors and actions toward achieving solutions to the joint problem. It does not consider whether specific actions are complied with, nor if the problem actually gets solved. These approaches to effectiveness definitions are an important means to conceptualize the variations in how effectiveness can be defined, which are adapted in a model (see Chapter 9) for evaluating effective transboundary water cooperation.

Despite these challenges and the variability present in definitions of effectiveness, there is agreement on how this problem of defining effectiveness is conceptualized. Both Underdal (1992, 2002) and Bernauer (1995) separately identify the three core questions which a framework studying regime effectiveness must address. If effectiveness is the resolution of a problem, essentially the success of the regime, then to evaluate the effectiveness, the progress must be compared to some standard of success. Therefore, Underdal (1992, 228–29) summarizes this issue as

> (i) What precisely constitutes *the object* to be evaluated? (ii) Against which *standard* is this object to be evaluated? (iii) *How* do we operationally go about comparing the object to our standard; in other words, what kind of measurement operations do we have to perform in order to attribute a certain *score* of "effectiveness to a certain object (regime)?"
>
> (emphasis in original)

Bernauer (1995, 335) similarly suggests,

> The concept of institutional effects raises three questions: Which outcomes do institutions affect and which of these outcomes should analysts focus on? How can these outcomes be evaluated in terms of institutional success or failure? Which measurement operations are required to assess the effect of an institution?

The phrasing of these two sets of questions represents the two approaches to research on effectiveness, explained in the previous paragraph. Underdal (1992) exemplifies the approach toward defining what effectiveness means as the dependent variable, while Bernauer's (1995) perspective approaches effectiveness through the analysis of the independent variables that explain whether and to what extent a regime is effective. This book is aiming to evaluate and define effectiveness and, therefore, aligns more closely with the questions prompted by Underdal (1992, 2002).

In terms of the object to be evaluated, the regime would be the obvious answer; however, there are distinct elements of the regime and regime formation process that will affect what is being evaluated and the data that is available. The formal *output* of a decision-making process is the "norms, principles, and rules constituting the regime itself" (Underdal 2002, 5). This differs from the results stemming from the implementation and adoption of this regime – or the *outcomes* and *impacts* (Underdal 2002). The *outcomes* are the changes in human behavior and result from "whether the operation of a regime causes one or more of its members … to alter their behavior, either by doing things they would not otherwise have done or by terminating or redirecting prior patterns of behavior" (Young 1994, 145). This is linked to compliance and the

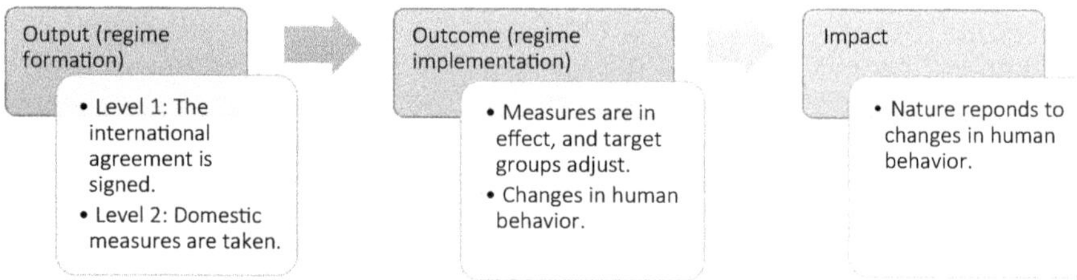

Figure 7.1 Stages of regime effectiveness including output, outcome (which differs in definition from the outcome frame), and impact.

Source: This figure is adapted from Underdal 2002 in Miles, Edward L., Steinar Andresen, Elaine M. Carlin, Jon Birger Skjærseth, Arild Underdal, and Jorgen Wettestad., *Environmental Regime Effectiveness*, p. 7, © 2001 Massachusetts Institute of Technology, by permission of The MIT Press.

extent that actors conform with provisions and mechanisms agreed upon in a treaty or other arrangement (R.B. Mitchell 1994). *Impacts*, on the other hand, are the physical changes in the environment – the ultimate goal or interest of the regime (Underdal 2002). Underdal (2002) proposes that these three elements are part of a chain of events and represent stages of the regime over time (see Figure 7.1). The *output* results from the end of the regime formation, then actors adjust their behavior, resulting in *outcomes*, which are followed by the *impact* on the bio-physical environment (Underdal 2002).

Determining the standard with which to evaluate the object has several considerations. First, within the *impact*, scholars have two predominant understandings of what effectiveness means – goal attainment or problem solving (Schmeier 2013). This will affect the standard that the *impact* is evaluated against. Goal attainment focuses on "the extent to which a regime's (stated or unstated) goals are attained over time" (Young 1994, 144; Schmeier 2013). Goals established by a regime can be vaguely defined or not feasible to measure, making them often impractical standards of evaluation (Schmeier 2013; Andresen, Skjaerseth, and Wettestad 1995). Alternatively, other scholars consider the standard to eliminate or alleviate the problem that the regime was established to solve (Young and Levy 1999). Casually, goal attainment and problem solving may seem identical; however, there is a clear distinction. Goals established by the regime may not solve or resolve the problem that spurred regime formation. The problem-solving approach remains focused on the original problem regardless of the goals established by the regime. In reality, this distinction may limit the interpretation of effectiveness, as it could consider the adaption of a regime to be ineffective. Therefore, I propose that both elements are needed for standards of evaluation.

Second, several scholars have developed a standard of evaluation based on points of reference to determine the relative improvement created by the regime or a comparison with the good or ideal solution (Underdal 2002).

The state of the regime can be compared to the collective optimum, which is the "one that accomplishes, for the group of members, all that can be accomplished – given the state of knowledge at the time" (Underdal 2002, 8). Alternatively, this could be compared with the counterfactual – either the no-regime counterfactual, which is the hypothetical state if there was no regime in place (Young 2011; Underdal 2002; 1992), or the baseline counterfactual, which is the hypothetical situation if the previous rules, norms, and principles had remained in place (Underdal 2002; 1992). Young (2011) proposes that effectiveness (E) is the difference of the actual performance (AP) from the no-regime counterfactual (NR) divided by the collective optimum (CO) from the no-regime counterfactual.

$$E = \frac{AP - NR}{CO - NR}$$

However, while useful in its representation of effectiveness, this equation is difficult to measure and quantify.

The final consideration plaguing the conceptualization of effectiveness, as discussed above, is the attribution of a certain score to a regime. This question is immensely complex, and research efforts seem to be focused on the previous two questions. Underdal (1992; 2002) proposes that for their research, the effectiveness will be accounted for at the ordinal level – or the success or failure of the regime in its problem solving along the distance to the collective optimum. Miles et al. (2002) present a discussion of operational measurement as well as a thorough discussion of the challenges, which are too extensive to be discussed here.

While effectiveness research has occurred most extensively in international relations, with respect to regime theory and international environmental politics, it has also been used within the international waters literature.

Effectiveness in the transboundary water literature

The terms effective and effectiveness are used regularly throughout the international and transboundary water literature. Like cooperation, effective is used in the common-sense use of the term – without defining it. Intuitively the reader is interpreting the conventional uses of the terms along the lines of the dictionary definition. Merriam-Webster defines effective as "producing a decided, decisive, or desired effect; impressive, striking; being in effect, operative; actual; or ready for service or action" (Merriam-Webster 2021). This conventional use of the term, again like cooperation, can make academically defining effectiveness in the context of key transboundary water concepts difficult, as the word is commonly used and understood. The use of effective and effectiveness has become increasingly more prevalent in international water literature over the last two decades. Much of it draws from the international relations and regime theory literature discussed above. Effective water governance, treaty effectiveness, effective management, effective water sharing, and effective river basin organizations (RBOs) or institutions are some of the commonly used terms, in addition to effective cooperation. The following Table 7.2 provides a few examples of definitions of effective or effectiveness in the context of common international water concepts. In structure, these definitions seem to follow the two general approaches to effectiveness

Table 7.2 Example of definitions of effective or effectiveness relative to standard terms in the international waters literature

Term	*Definition or description*	*Source*
Enhancing the effectiveness of water governance	Principle 1: "Clearly allocate and distinguish *roles and responsibilities* for water policymaking, policy implementation, operational management and regulation, and foster co-ordination across these responsible authorities." Principle 2: "Manage water at the *appropriate scale(s)* within integrated basin governance systems to reflect local conditions, and foster co-ordination between the different scales." Principle 3: "Encourage policy coherence through effective *cross-sectoral co-ordination*, especially between policies for water and the environment, health, energy, agriculture, industry, spatial planning, and land use." Principle 4: "Adapt the level of capacity of responsible authorities to the complexity of water challenges to be met, and to the set of competencies required to carry out their duties."	12 OECD Principles on Water Governance (OECD 2015, 9–10)
Treaty effectiveness or success	"The effectiveness or success of these treaties can be assessed in at least two ways: (1) the degree to which a treaty resolves a specific issue that it is designed to solve, and (2) the degree to which it increases subsequent water cooperation more generally."	(Brochmann 2012, 144)
RBO effectiveness	RBO efficiency and effectiveness are dependent on: • "A treaty and agreement for cooperation." • "The principle of water sharing." • "The principle of integrated planning." • "An adaptable management structure." • "A conflict resolution mechanism."	(Komakech, Jaspers, and van Der Zaag 2007; Do Ó 2012, 785)

Table 7.2 Cont.

Term	*Definition or description*	*Source*
Treaty effectiveness	"In order to analyze the operation and effectiveness of these treaties and MoUs, this article will look at five core elements used to evaluate the legal aspects of transboundary water regimes: scope, substantive rules, procedural rules, institutional mechanisms, and dispute settlement mechanisms."	(Liu 2015, 360)
Effectiveness of water governance	"Effectiveness of water governance relates to the contribution of governance to defining clear sustainable water policy goals and targets at different levels of government, to implement those policy goals, and to meet expected objectives or targets."	(OECD 2018, 5)
Effective transboundary water sharing	"Efficient and effective transboundary water sharing should be based on four guiding principles: coordination and cooperation, interdisciplinary analysis, watershed, and river basin planning, and adaptive management."	(Environmental and Water Resources Institute 2012, 2)
Effective water management	"Effective water management relies on a wide range of institutions and actors playing distinct but inter-connected roles. Coordination and cooperation are essential to ensure effective water management, health protection, and sustainable development. Good water policies implemented by nationally tailored water legislation and other tools can facilitate coordination and help governments achieve their water management objectives."	(FAO Legal Office 2009, 16)

research in international relations. Most of the definitions listed in Table 7.2 are approaching effectiveness by attempting to determine the independent variables that establish the extent of effectiveness, while the example from Brochmann (2012) attempts to define treaty effectiveness. This definition also incorporates problem solving and an overall goal of increasing cooperation (Brochmann 2012). In order to look more closely at how effectiveness is used in the water literature, the following subsections present three examples defining and/or measuring effective water governance, treaty effectiveness, and RBO effectiveness. The goal of examining several examples of effectiveness in the water literature is to see if there are common structures or examples that we can draw from when developing the definition and method of evaluation for effective transboundary water cooperation.

Effective water governance

In 2003, the Global Water Partnership (GWP) commissioned a Technical Background Paper, *Effective Water Governance*, that aimed to present a discussion of governance and how it is relevant to the water management sector (Rogers and Hall 2003). As an organization, the GWP is tasked with the promotion and knowledge development of integrated water resources management (IWRM). This paper aimed to increase knowledge and understanding of governance in the water sector in order to support the changes and policies required to support IWRM. This discussion of water governance has an emphasis on domestic waters but does not exclude internationally shared waters. Despite the title, the main emphasis of the paper is not on effectiveness, but rather it begins with a discussion of governance and the particulars of water governance. The authors define water governance as "the range of political, social, economic, and administrative systems that are in place to develop and manage water resources, and the delivery of water services at different levels of society" (Rogers and Hall 2003). This is one of the most frequently cited definitions of water governance, and this paper introduces the concept of 'effective water governance' (Villholth et al. 2018). Rogers and Hall (2003) argue that it is important for attributes that make water governance effective to be identified; however, they recognize that there is 'no single model of effective governance.' In order for governance to be effective, a governance system must be reflective of context – including the social, economic, and cultural factors of a basin, aquifer, or state.

Ultimately, the authors provide a descriptive list of attributes that are considered central to having effective water governance. These are separated into two groups – approaches to governance and the performance and operation of governance systems (Rogers and Hall 2003, 26–29):

Approaches to effective governance

- *Open and transparent*: For a governance system to be effective, its institutions should work in an open and transparent manner. Policy decisions should be transparent to both actors in the process as well as stakeholders and the greater public.
- *Inclusive and communicative*: All levels of government should be inclusive with respect to participation in the development and implementation of policies.
- *Coherent and integrative*: Policies should be coherent and integrate across borders and sectors. In addition, governance should integrate traditional users and their needs in policy with the other potential users and sectors that have previously been separate.

- *Equitable and ethical*: There needs to be equity between various users and interests. The principles of equity should be context specific.

Performance and operation for effective governance systems

- *Accountable*: Actors within the governance system, including government, private sector, and civil society, need to be held accountable to the public and institutional stakeholders.
- *Efficient*: Governance systems should minimize transaction costs to achieve political and economic efficiency; however, social and environmental efficiency needs to be balanced.
- *Responsive and sustainable*: Governance should consider future needs and long-term sustainability while still meeting the needs of the present in a responsive and proportionate manner at the appropriate scale.

These principles for effective governance do not clearly define what effective water governance is but rather present behavioral changes that would need to take place to have the potential for effective governance. Given the normative nature of many of these principles, it would be difficult to determine, measure, and even evaluate the extent of effectiveness of the governance. However, they do allow for states and other actors to have the flexibility to establish the institutions, arrangements, norms, and other rules and policies that are applicable to their context while encouraging the incorporation of these 'good governance' principles. Given its construction, effective water governance is often used interchangeably with 'good water governance.'

Treaty effectiveness

From the literature and interview data I collected, the most commonly used term in conjunction with the idea of effectiveness, other than effective cooperation, is treaty effectiveness or effectiveness of agreements. Treaties, agreements, and other international arrangements are extensively studied in the international water literature and are often used as proxies for measuring cooperation. It follows that the effectiveness of treaties is also measured and, in many cases, they are used as proxies or components in the determination of the effectiveness of cooperation. Of the 273 articles within the sample, 8 directly mention, define, and/or measure treaty effectiveness. However, there are substantially more articles and reports that evaluate treaty effectiveness, explicitly or implicitly, than discussed in this sub-section (for examples, see De Stefano et al. 2012; Drieschova, Giordano, and Fischhendler 2008; Stinnett and Tir 2009; Tir and Stinnett 2012; Brochmann 2012; Brochmann, Hensel, and Tir 2012; Dinar et al. 2015; Choudhury and Islam 2015; S.M. Mitchell and Zawahri 2015). Even within this relatively narrow scope of a treaty, as compared to cooperation and governance, the interpretation of what makes a treaty effective varies. For example, De Stefano et al. (2017), in their larger study of hydropolitical tensions in transboundary river basins, provide a succinct summary of the literature that has identified the characteristics that improve treaty effectiveness. These include "flexible management structure, clear and flexible allocating criteria, equitable distribution of benefits, detailed conflict resolution mechanisms (Giordano and Wolf 2003), and mechanisms for increasing resilience towards water variability (Drieschova, Giordano, and Fischhendler 2008; De Stefano et al. 2012), such as flexible but specific water allocation mechanisms (Dinar et al. 2015). Institutional models should also make explicit provisions for institutional learning and change (Meinzen-Dick 2007). Additionally, treaties that include a direct enforcement measure, an adaptability mechanism, and a self-enforcement clause present higher levels of cooperation relative to those that do not have such provisions (Dinar et al. 2015)" (De Stefano et al. 2017, 37).

This differs from Liu (2015) – they analyze the cooperation in Sino-Indian waters based on the effectiveness of the treaties and MOUs. The core elements they use are adapted from Wouters (2013) and evaluate the legal regime by considering: scope, substantive rules, procedural rules, institutional mechanisms, and dispute settlement mechanisms (Liu 2015, 360). While there are similarities, the specificity of the characteristics identified in De Stefano et al. (2017) compared with the broad yet overlapping categories used by Liu (2015) would result in highly different interpretations of whether and to what extent a treaty is effective.

Several other articles take a different approach by defining treaty effectiveness explicitly, rather than working to identify the characteristics of effective treaties. Brochmann et al. (2012) consider how the origins and conditions under which a treaty is signed contribute to its effectiveness. They define treaty effectiveness as preventing future disputes between the riparians. Specifically, this study determines if the presence of an agreement, the pre-conditions for signing, and the broad relationship between the riparians impacts whether a treaty is more effective at preventing or reducing water-related disputes (Brochmann, Hensel, and Tir 2012). Mitchell and Zawahri (2015) define effectiveness of treaty design in a similar fashion. They argue that a treaty or institution is effective if it is able to peacefully address disputes between riparians. Treaties will be more successful (or effective) if they are supported by RBOs and include mechanisms for information exchange, monitoring, enforcement, and conflict resolution (S.M. Mitchell and Zawahri 2015). These two examples identify a definition of effective treaties – resolution or prevention of disputes – that provide a standard with which to evaluate the treaty design or external variables to treaty establishment. However, in these and similar studies it is difficult, if not impossible, to determine causality.

River basin organization and/or institutional effectiveness

Evaluating and conceptualizing RBOs and/or institutional effectiveness is also common. Within the sample collected, six articles mentioned, defined, and/or measured institutional or RBO effectiveness at various levels of detail. Like treaties, cooperative institutional frameworks – which can include treaties and institutions – and RBOs, specifically, are often studied for their role in establishing and maintaining transboundary water cooperation. The presence of an RBO, for example, is often used in measures of cooperation or hydropolitical resilience (e.g., De Stefano et al. 2017; 2012). Much of the discussion on effective institutions is focused on institutional design, or it is discussed in the abstract, by calling for effective institutions to solve transboundary water issues. For example, Albrecht et al. (2018, 1101) highlight effective institutions without defining them or identifying characteristics: "Effective institutional arrangements are often invoked to address transboundary water issues, such as conflict resolution, risk management, and trade-off negotiation (Pahl-Wostl, Palmer, and Richards 2013)." However, several scholars have developed models for institutional or RBO effectiveness. Berardo and Gerlak (2012, 103) present a "multi-faceted definition of institutional effectiveness that speaks to the broad capacity of the institution to perform the tasks for which it was designed." This model consists of two levels; the first level examines the interstate agreement, which structures the relationship between the riparians. This includes "process design, which refers to the complex internal negotiations that usually take place inside new institutions to decide how, when, and under what conditions the parties take place" (Berardo and Gerlak 2012, 104). The second level addresses the elements of the process design. They identify four features that will contribute to collaborative solutions: transparency in the decision-making process, the production and dissemination of scientific knowledge, conflict resolution, and public input and representation (Berardo and Gerlak 2012). These aspects of effectiveness are measured against the desired outcome to reach collaborative solutions to problems within the basin. This model defines the object to be evaluated in a broad sense that incorporates agreements as well as institutions, such as RBOs. Other scholars focus more closely on the role of RBOs in effective cooperation.

Schmeier has extensively and comprehensively evaluated RBOs and their role in transboundary water governance, including evaluating their effectiveness. She defines an RBO as effective "if it provides principles, norms, and rules, and performs activities that contribute to behaviour changes among riparian actors (outcome) leading to the effective governance of the river basin (impact)" (Schmeier 2012, 14). This definition is expanded to defining RBO effectiveness as "the extent to which an RBO contributes to behavior changes among riparian actors, ultimately contributing to the solution of the collective action problem that prompted the RBO's establishment and the promotion of joint governance of water-related collective action problems in the basin" (Schmeier 2013, 26). This definition is broken into three dimensions of effectiveness which aim to capture the complexity in evaluating effectiveness. The first dimension is the effectiveness level, which separates the outcome and impact of effectiveness. Effectiveness scope is the second dimension, which captures potential issues that an RBO would contribute to, including political stability, environmental stability, economic growth, and social development.

Lastly is the effectiveness range, which encompasses the extent to which RBOs contribute to water collective action problems and the extent they contribute to broader non-water goals within the river basin (Schmeier 2013). Using this understanding of RBO effectiveness, she identifies the independent variables that can be causally attributed to RBO effectiveness. These include the institutional design, the problem structure, and the situation structure. Within these, she identifies and hypothesizes the impacts of various elements and explanatory factors, which are too detailed to discuss here, see Schmeier 2012; 2013. Overall, her research into RBO effectiveness shows that certain institutional design elements, such as data and information sharing, dispute settlement mechanisms, and sustainable funding, significantly impact RBO effectiveness; the nature of the problem as well as the constellation of the actors – problem- and situation structure, respectively – can also affect the likelihood of effective river basin governance (Schmeier 2012; 2013).

Of these examples and the descriptions in Table 7.2, effectiveness seems to be more clearly defined and modeled for treaties and institutions, than for governance and management. Given the difficulty and disagreement in defining governance, management, cooperation, and diplomacy, it is logical that treaties and RBOs with arguably more identifiable mechanisms and design elements are evaluated for effectiveness. In addition, these are key aspects, and often proxies, for transboundary water cooperation. Despite this, there has been limited research on defining effectiveness in the context of transboundary water cooperation. The phrase 'effective cooperation,' or some variation, is commonly found in the literature but, generally, it is not defined. There have been increasing calls for effective cooperation, successful cooperation, or, comparably, operational cooperation. For example, Tarlock (2015) argues that the focus should be on promoting effective cooperation, where the objective is to produce measurable benefits – or output cooperation. Only two prior studies in the water literature have addressed effective transboundary water cooperation by developing a definition and/or model for evaluation.

Foundational studies on effective transboundary water cooperation

The Hague Institute for Global Justice, in conjunction with partner organizations, produced the report *The Multi-track Water Diplomacy Framework: A Legal and Political Economy Analysis for Advancing Cooperation over Shared Waters* as a part of the Water Diplomacy: Making Water Cooperation Work project. The main research question for this report is "What are the key factors affecting transboundary water cooperation?" (Huntjens et al. 2016, 5). Their aim in answering this question is to determine the factors that influence cooperation and ultimately improve water diplomacy. The paper results in a Multi-Track Water Diplomacy Framework, which aims to identify water problems by identifying intervention points, while proposing sustainable solutions. (see Huntjens et al. 2016 for full description and graphical representation of the framework.) This framework incorporates the factors identified as affecting cooperation, which are the characteristics that compose effective cooperation. The authors conduct a review of concepts in the literature that are associated with effective cooperation; these include: "mutual satisfaction, the role of negotiation and procedural and outcome based justice, the sustainability and benefits of cooperation, the relations between peace and conflict and cooperation, the role of trust and reciprocity, information sharing, policy learning, adaptive governance, and multi-level governance" (Huntjens et al. 2016). Based on their literature review of these topics, they propose a working definition of effective water cooperation: "a collaboration in which two or more parties find a negotiated compromise on maximizing mutual gains and achieving joint wins for all parties involved, resulting in the availability of an acceptable quantity and quality of water for health, livelihoods, ecosystems and production, coupled with an acceptable level of water-related risks to people, environments, and economies" (Huntjens et al. 2016, 19).

This definition is essentially arguing that effective cooperation is collaboration through negotiation for mutual gains, which include gains in water security. The end clause of the definition is similar to the definition of water security defined by Grey and Sadoff (2007, 545). These authors have essentially argued for the standard of comparison for effective cooperation to be water security, which also coincides with Tarlock's (2015) argument that effective cooperation should produce outcomes that promote and improve water security, as mentioned above.

The factors identified are incorporated into the framework for water diplomacy in the "zone of possible effective cooperation" (Huntjens et al. 2016). This is based on the idea of the zone of possible agreement in negotiation theory and represents the areas that have the potential to promote effective cooperation in a basin. These are future areas where cooperation can occur and be effective; they can be used to identify high and low priorities for future cooperation and be a mechanism to encourage a shift toward mutual gains in negotiations (Huntjens et al. 2016).

The second study that considers the definition and evaluation of effective transboundary water cooperation is by Daoudy and Kistin (2008) – *Beyond Water Conflict: Evaluating the Effects of International Water Cooperation*. This paper addresses the research questions: "How can the consequences of transboundary water cooperation be assessed? How can effective water cooperation be defined and measured?" (Daoudy and Kistin 2008, 3). After a review of the literature on cooperation, they define transboundary water cooperation – the independent variable – as " an on-going and non-linear process in which state and non-state actors establish challenge, modify and legitimize multi-layered governance structures" that encompasses more than the treaties and agreements (Daoudy and Kistin 2008, 10). Using this definition of cooperation, they propose an analytical framework on which they base their effectiveness evaluation; it includes elements of power, negotiation, agreements, actor behavior and interests, and implementation/reception (see figure 3 in Daoudy and Kistin 2008). Effectiveness evaluation is generally asking, "is it executing its designated functions? Producing results? Living up to expectations?" (Daoudy and Kistin 2008, 11). However, as discussed in the previous sections, it is challenging to devise criteria to evaluate the effectiveness, particularly in a comparable way. The authors' literature review highlights the most common approaches to measuring effectiveness: compliance, goal achievement, and problem solving (Daoudy and Kistin 2008). To evaluate compliance, the extent to which the actor's behavior conformed to the rules must be measured, such as whether specified water allocations are completed. Goal attainment is the extent to which partners fulfill the joint objectives established by the parties, which may differ from individual state interests. Therefore, they also propose a variety based on interest achievement, which considers whether actors' interests have been met and how power influences whose interests are achieved. Problem solving is evaluated based on whether or not cooperative efforts solve problems as they were intended to – which is often evaluated on a binary scale of success or failure to solve the problem (Daoudy and Kistin 2008). These discussions are drawn primarily from the literature on regime effectiveness within international relations. Lastly, the authors point to literature that acknowledges that there are generally two strategies for measuring these criteria of effectiveness: inferences and indices (Young 2001; Daoudy and Kistin 2008). Inferences aim to establish causal links between the regime and the solution, whereas indices are generally employed to provide a standard measure that is comparable between regimes.

Based on this literature review of effectiveness and the analytical framework of cooperation, the authors evaluate the effectiveness of the cooperation in the Jordan and Tigris-Euphrates River Basins. Despite the definition of cooperation acknowledging that the cooperation is more than the agreements, the analysis is focused on the agreements signed between the riparian states; however, the power structure is taken into consideration when evaluating whether parties' interests have been met. The authors evaluated the cooperation based on compliance, goal attainment, interest achievement, and problem solving, which, interestingly, yields mixed results depending on the agreement. For example, Daoudy and Kistin (2008) evaluate the *Water Protocol between Syria and Turkey* signed in June 2002. The focus of the agreement is on the exchange of expertise and mutual training; the authors cite that the terms are complied with and the goals identified have been attained; however, the problem that necessitated the protocol was not solved, as there was "no long-term allocation of common waters on [the] basis of equitable and reasonable utilization" (Daoudy and Kistin 2008, 21).

Conclusion: Moving from literature to data

This study by Daoudy and Kistin (2008) and the report *The Multi-track Water Diplomacy Framework* by Huntjens et al. (2016) both qualitatively identify elements of cooperation and effectiveness based on the international water and the international relations literature. The results are two different perspectives on effectiveness, based on the desired goals of the study. Huntjens et al. (2016) actually develop a definition of effectiveness, which is based on a standard of water security. However, despite having this standard, cooperation is not evaluated against whether water security has been achieved in the basin, but rather the idea is used to identify potential actions or solutions within the zone of effective cooperation to improve negotiation and overall water diplomacy in the basin. Comparatively, Daoudy and Kistin (2008) do not define effective cooperation in a concise definition, nor a specific standard of evaluation. Rather, they present four alternative criteria for evaluating effectiveness, where each criterion is context specific to the cooperative agreement under evaluation. Part III builds on the research presented in these two studies on effective transboundary water cooperation, as well as general literature on effectiveness and cooperation.

Now that we have discussed the current thinking on what is or is not effective in the context of regimes, institutions, treaties, governance, and cooperation, the next chapter will shift focus. This will approach Part III by considering the alternative path for answering the second challenge – how to develop a definition of effective transboundary water cooperation based on the sample literature and interview data.

Note

1 Please keep in mind when reading that this chapter focuses on defining *transboundary water cooperation*, not cooperation in the broader sense of the word. I use *cooperation* without the *transboundary water* modifier; however, I am referring to *transboundary water cooperation*. I will note when I refer to cooperation unlinked to transboundary water.

References

Albrecht, Tamee R., Robert G. Varady, Adriana A. Zuniga-Teran, Andrea K. Gerlak, Rafael Routson De Grenade, América Lutz-Ley, Facundo Martín, et al. 2018. "Unraveling Transboundary Water Security in the Arid Americas." *Water International* 43 (8): 1075–1113. https://doi.org/10.1080/02508060.2018.1541583.

Andresen, Steinar, Jon Birger Skjaerseth, and Jørgen Wettestad. 1995. "Regime, the State and Society: Analyzing the Implementation of International Environmental Commitments." *International Institute for Applied Systems Analysis*, 75.

Berardo, Ramiro, and Andrea K. Gerlak. 2012. "Conflict and Cooperation along International Rivers: Crafting a Model of Institutional Effectiveness." *Global Environmental Politics* 12 (1): 101–20.

Bernauer, Thomas. 1995. "The Effect of International Environmental Institutions: How We Might Learn More." *International Organization* 49 (2): 351–77. https://doi.org/10.1017/S0020818300028423.

———. 2002. "Explaining Success and Failure in International River Management." *Aquatic Sciences* 64 (1): 1–19. https://doi.org/10.1007/s00027-002-8050-4.

Brochmann, Marit. 2012. "Signing River Treaties—Does It Improve River Cooperation?" *International Interactions* 38 (2): 141–63. https://doi.org/10.1080/03050629.2012.657575.

Brochmann, Marit, Paul R. Hensel, and Jaroslav Tir. 2012. "International River Treaty Effectiveness." 23–26. San Diego. http://citeseerx.ist.psu.edu/viewdoc/download?doi=10.1.1.727.2938&rep=rep1&type=pdf.

Choudhury, Enamul, and Shafiqul Islam. 2015. "Nature of Transboundary Water Conflicts: Issues of Complexity and the Enabling Conditions for Negotiated Cooperation." *Journal of Contemporary Water Research & Education* 155 (1): 43–52. https://doi.org/10.1111/j.1936-704X.2015.03194.x.

Daoudy, Marwa, and Elizabeth J. Kistin. 2008. "Beyond Water Conflict: Evaluating the Effects of International Water Cooperation." San Francisco.

De Stefano, Lucia, James Duncan, Shlomi Dinar, Kerstin Stahl, Kenneth M Strzepek, and Aaron T Wolf. 2012. "Climate Change and the Institutional Resilience of International River Basins." *Journal of Peace Research* 49 (1): 193–209.

De Stefano, Lucia, Jacob D. Petersen-Perlman, Eric A. Sproles, Jim Eynard, and Aaron T. Wolf. 2017. "Assessment of Transboundary River Basins for Potential Hydro-Political Tensions." *Global Environmental Change* 45 (July): 35–46. https://doi.org/10.1016/j.gloenvcha.2017.04.008.

Dinar, Shlomi, David Katz, Lucia De Stefano, and Brian Blankespoor. 2015. "Climate Change, Conflict, and Cooperation: Global Analysis of the Effectiveness of International River Treaties in Addressing Water Variability." *Political Geography* 45 (March): 55–66. https://doi.org/10.1016/j.polgeo.2014.08.003.

Do Ó, Afonso. 2012. "Drought Planning and Management in Transboundary River Basins: The Case of the Iberian Guadiana." *Water Policy* 14 (5): 784–99. https://doi.org/10.2166/wp.2012.173.

Drieschova, Alena, Mark Giordano, and Itay Fischhendler. 2008. "Governance Mechanisms to Address Flow Variability in Water Treaties." *Global Environmental Change* 18 (2): 285–95. https://doi.org/10.1016/j.gloenvcha.2008.01.005.

Environmental and Water Resources Institute. 2012. *Guidelines for Development of Effective Water Sharing Agreements*. Reston, Virginia: American Society of Civil Engineers.

FAO Legal Office. 2009. "Law for Water Management: A Guide to Concepts and Effective Approaches." FAO Legislative Study 101. Rome: FAO.

Giordano, Meredith A., and Aaron T. Wolf. 2003. "Sharing Waters: Post-Rio International Water Management." *Natural Resources Forum* 27: 163–71.

Grey, David, and Claudia W. Sadoff. 2007. "Sink or Swim? Water Security for Growth and Development." *Water Policy* 9 (September): 545. https://doi.org/10.2166/wp.2007.021.

Helm, Carsten, and Detlef Sprinz. 2000. "Measuring the Effectiveness of International Environmental Regimes." *Journal of Conflict Resolution* 44 (5): 630–52. https://doi.org/10.1177/0022002700044005004.

Huntjens, Patrick, Yumiko Yasuda, Ashok Swain, Rens de Man, Bjørn-Oliver Magsig, and Shafiqul Islam. 2016. "The Multi-Track Water Diplomacy Framework: A Legel and Political Economy Analysis for Advancing Cooperation over Shared Waters." Hague: The Hague Institute for Global Justice. www.thehagueinstituteforglobaljustice.org/thinking-doing/publication/the-multi-track-water-diplomacy-framework-a-legal-and-political-economy-analysis-for-advancing-cooperation-over-shared-waters/.

Kay, David A., and Harold K Jacobson. 1983. *Environmental Protection: The International Dimension*. Totowa, NJ: Osmun.

Keohane, Robert O., and Lisa L. Martin. 1995. "The Promise of Institutionalist Theory." *International Security* 20 (1): 39. https://doi.org/10.2307/2539214.

Kistin, Elizabeth J. 2007. "Trans-Boundary Cooperation in SADC: From Concept to Implementation." In *8th WaterNet/WARFSA/GWP-SA Symposium, Lusaka, Zambia*. Vol. 30.

Komakech, Charles H., Frank Jaspers, and Pieter van Der Zaag. 2007. "Strengthening Transboundary River Basin Institutions: A Comparative Learning from Basin Institutions in Africa, Europe and Southeast Asia." In *Reducing the Vulnerability of Societies to Water Related Risks at the Basin Scale*, edited by A. Schumann and M. Pahlow, 157–64. Bochum, Germany: International Association of Hydrological Sciences.

Kranz, Nicole, Timo Menniken, and Jochen Hinkel. 2010. "Climate Change Adaptation Strategies in the Mekong and Orange-Senqu Basins: What Determines the State-of-Play?" *Environmental Science & Policy* 13 (7): 648–59. https://doi.org/10.1016/j.envsci.2010.09.003.

Krasner, Stephan D. 1983. "Structural Causes and Regime Consequences: Regimes as Intervening Variables." In *International Regimes*, edited by S.D. Krasner, 1–21. London: Ithaca.

Levy, Marc A., Oran R. Young, and Michael Zürn. 1995. "The Study of International Regimes." *European Journal of International Relations* 1 (3): 267–330. https://doi.org/10.1177/1354066195001003001.

Levy, Marc, Robert O. Keohane, and Peter M. Haas. 1993. "Improving the Effectiveness of International Environmental Institutions." In *Institutions for the Earth: Sources of Effective International Environmental Protection*, edited by Peter M. Haas, Robert O. Robert Owen Keohane, and Marc A. Levy, 397–426. Global Environmental Accords Series. Cambridge, Mass.: MIT Press.

Lindemann, Stefan. 2008. "Understanding Water Regime Formation—A Research Framework with Lessons from Europe." *Global Environmental Politics* 8 (4): 117–40. https://doi.org/10.1162/glep.2008.8.4.117.

Liu, Yang. 2015. "Transboundary Water Cooperation on the Yarlung Zangbo/Brahmaputra – a Legal Analysis of Riparian State Practice." *Water International* 40 (2): 354–74. https://doi.org/10.1080/02508060.2015.1015319.

Meinzen-Dick, R. 2007. "Beyond Panaceas in Water Institutions." *Proceedings of the National Academy of Sciences* 104 (39): 15200–205. https://doi.org/10.1073/pnas.0702296104.

Merriam-Webster. 2021. "Effective." In. www.merriam-webster.com/dictionary/effective.

Miles, Edward, Arild Underdal, Steinar Andresen, Jørgen Wettestad, Jon Birger. Skjaerseth, and Elaine Carlin, eds. 2002. *Environmental Regime Effectiveness: Confronting Theory with Evidence*. Cambridge, MA: MIT Press.

Mitchell, Ronald B. 1994. "Regime Design Matters: Intentional Oil Pollution and Treaty Compliance." *International Organization* 48 (3): 425–58. https://doi.org/10.1017/S0020818300028253.

Mitchell, Sara McLaughlin, and Neda A. Zawahri. 2015. "The Effectiveness of Treaty Design in Addressing Water Disputes." *Journal of Peace Research* 52 (2): 187–200. https://doi.org/10.1177/0022343314559623.

OECD. 2015. "OECD Principles on Water Governance." Paris: Organisation for Economic Co-operation and Development. www.oecd.org/cfe/regional-policy/OECD-Principles-on-Water-Governance-brochure.pdf.

———. 2018. "OECD Water Governance Indicator Framework." OECD Water Governance Programme. www.oecd-ilibrary.org/environment/implementing-the-oecd-principles-on-water-governance/oecd-water-governance-indicator-framework_9789264292659-5-en.

O'Neill, Kate, Jörg Balsiger, and Stacy D. VanDeveer. 2004. "ACTORS, NORMS, AND IMPACT: Recent International Cooperation Theory and the Influence of the Agent-Structure Debate." *Annual Review of Political Science* 7 (1): 149–75. https://doi.org/10.1146/annurev.polisci.7.090803.161821.

Pahl-Wostl, Claudia, Margaret Palmer, and Keith Richards. 2013. "Enhancing Water Security for the Benefits of Humans and Nature—the Role of Governance." *Current Opinion in Environmental Sustainability* 5 (6): 676–84. https://doi.org/10.1016/j.cosust.2013.10.018.

Phillips, David, Marwa Daoudy, Stephen McCaffrey, Joakim Öjendal, and Anthony Turton. 2006. *Trans-Boundary Water Cooperation as a Tool for Conflict Prevention and Broader Benefit-Sharing*. Ministry for Foreign Affairs Stockholm.

Rogers, Peter, and Alan W. Hall. 2003. *Effective Water Governance*. Technical Committee Background Papers 7. Stockholm: Global Water Partnership.

Schmeier, Susanne. 2012. "Navigating Cooperation beyond the Absence of Conflict: Mapping Determinants for the Effectiveness of River Basin Organisations." *International Journal of Sustainable Society* 4 (1–2): 11–27.

———. 2013. *Governing International Watercourses: River Basin Organizations and the Sustainable Governance of Internationally Shared Rivers and Lakes*. New York: Routledge. www.routledge.com/Governing-International-Watercourses-River-Basin-Organizations-and-the/Schmeier/p/book/9780415623582.

Stinnett, Douglas M., and Jaroslav Tir. 2009. "The Institutionalization of River Treaties." *International Negotiation* 14 (2): 229–51. https://doi.org/10.1163/157180609X432815.

Tarlock, Dan. 2015. "Promoting Effective Water Management Cooperation Among Riparian Nations." *TEC Background Papers*, no. 21. www.gwp.org/Global/ToolBox/Publications/Background%20papers/GWP_TEC21_web.pdf.

Tir, Jaroslav, and Douglas M Stinnett. 2012. "Weathering Climate Change: Can Institutions Mitigate International Water Conflict?" *Journal of Peace Research* 49 (1): 211–25. https://doi.org/10.1177/0022343311427066.

Underdal, Arild. 1992. "The Concept of Regime 'Effectiveness.'" *Cooperation and Conflict* 27 (3): 227–40.

———. 2002. "One Question, Two Answers." In *Environmental Regime Effectiveness: Confronting Theory with Evidence*, edited by Edward Miles, Arild Underdal, Steinar Andresen, Jørgen Wettestad, Jon Birger Skjaerseth, and Elaine Carlin, 3–43. Cambridge, MA: MIT Press.

Villholth, Karen G., Elena López-Gunn, Kirstin Conti, Alberto Garrido, and Jac A. M. van der Gun, eds. 2018. *Advances in Groundwater Governance*. Leiden, The Netherlands: CRC Press/Balkema.

Wouters, Patricia. 2013. "International Law–Facilitating Transboundary Water Cooperation." *Published by the Global Water Partnership, TEC Background Papers*, no. 17. http://papers.ssrn.com/sol3/papers.cfm?abstract_id=2363809.

Young, Oran R. 1989. "The Politics of International Regime Formation: Managing Natural Resources and the Environment." *International Organization* 43 (3): 349–75.

———. 1994. "The Effectiveness of International Governance Systems." In *International Governance: Protecting the Environment in a Stateless Society*, by Oran R. Young, 140–60. Ithaca: Cornell University Press.

————. 1999. "Regime Effectiveness: Taking Stock." In *The Effectiveness of International Environmental Regimes: Causal Connections and Behavioral Mechanisms*, edited by Oran R. Young, 250–79. Cambridge, MA: The MIT Press.

———. 2001. "Inferences and Indices: Evaluating the Effectiveness of International Environmental Regimes." *Global Environmental Politics* 1 (1): 99–121. https://doi.org/10.1162/152638001570651.

———. 2011. "Effectiveness of International Environmental Regimes: Existing Knowledge, Cutting-Edge Themes, and Research Strategies." *Proceedings of the National Academy of Sciences* 108 (50): 19853–60.

Young, Oran R., and Marc Levy. 1999. "The Effectiveness of International Environmental Regimes." In *The Effectiveness of International Environmental Regimes: Causal Connections and Behavioral Mechanisms*, edited by Oran R. Young, 1–32. Cambridge, MA: The MIT Press.

8 Unpacking effectiveness

Developing a definition of effective transboundary water cooperation

In Part III, we are exploring the question: What makes transboundary water cooperation effective? Chapter 7 began exploring this question by reviewing the definitions of effectiveness and effective transboundary water cooperation in the literature. But in order to answer this question, we must also ask: What are we evaluating, what is the standard of comparison, and what biases and perspectives are represented in the definition? Ultimately, defining effective cooperation[1] is a subjective task. This chapter considers these questions in order to develop a consolidated understanding and definition of effective cooperation. To develop this definition, this chapter identifies the core components – or aspects of effectiveness – using a qualitative content analysis of expert[2] interviews, peer-reviewed articles, and grey literature collected and conducted by the author. Using these data aims to reduce the inherent subjectivity and incorporate multiple and diverse perspectives on what makes cooperation effective. These data were categorized into both predetermined and organically determined categories and themes – referred to as *codes*. The *codes* identified include implicit and explicit definitions of effective cooperation and other related terms, aspects of effective cooperation, methods to measure effectiveness, scale, barriers, time, and comparability.

Within the literature sample (n = 273) and interviews (n = 54), I identified 5 definitions of effective cooperation and 54 direct answers to the questions: 'How would you define effective transboundary water cooperation? What do you mean by effective?' Clearly stated, concise definitions of effective cooperation in the literature were not common. Compared to the question asking the interviewees to define cooperation, the interviewees were freer and more definitive in their discussions of how they define and perceive effective cooperation. Definitions varied depending on perspective, bias, and context; however, there were notable themes. As with the data on cooperation, I was able to identify aspects of effectiveness – or a word, phrase, or sentence(s) that represents a standalone concept of effective cooperation. A complex phrase or sentence may capture more than one aspect of effectiveness and, therefore, could be coded more than once. Many aspects of effectiveness overlap or have similarities with the key elements identified in the cooperation data analysis (see Chapter 5); however, I aimed to only code aspects of effectiveness when it was

DOI: 10.4324/9781003126249-11

clearly referencing effective cooperation or effectiveness. In total, I identified 45 aspects of effectiveness. Aspects include concepts such as achieving or defining outcomes, participation, stability, or information and data. For a complete list of the aspects of effectiveness collected, see Appendix B. The chapter aims to present a discussion of the data on how effectiveness and effective cooperation are defined in the literature and by academics and practitioners; it presents the main themes from the aspects of effectiveness, rather than an extensive discussion on each one. There are significant challenges to defining effectiveness and applying that to effective cooperation. This research is neither attempting to nor claiming to establish causality between the aspects of effectiveness and the success or failure of a cooperative process.

Explicit and implicit discussions of effective cooperation

It should come as no surprise to the reader that effective/effectiveness is a challenging concept to define and is variable based on what is being evaluated as effective, the context, and subjectivity of the person or body defining the standard for what is considered effective. Despite this, I still encountered definitions and discussions of effective cooperation in the literature. The data was separated into explicit and implicit discussions of effectiveness. I define explicit discussions of effectiveness as the article text or the interview transcript that directly discusses a framework for effectiveness, a method to measure effectiveness, or directly defined the term, either generally or with respect to cooperation, governance, treaty, institutional, or management. This was further refined into text that only defined effective cooperation. Implicit discussions of effectiveness are the text that addresses effectiveness indirectly, such as text that mentions that some aspects of cooperation add to its effectiveness. As with the explicit discussions, the implicit text was subdivided into those that reference effectiveness generally, with respect to a related term, and the text that was indirectly discussing effective cooperation. Within the data, there are only 18 implicit discussions of effective cooperation in the literature out of 52 implicit discussions of effectiveness, generally, within the entire sample of 273 articles and grey literature. A sample of the definitions is included in Table 8.1.

As with the implicit discussions of cooperation, effective cooperation is not defined in any of these quotes. Other terms that are often used interchangeably with effective cooperation are exemplified, such as 'successful transboundary cooperation.' The success or failure of cooperation is similar to the idea of effective or ineffective cooperation, and through the literature, they are often used interchangeably or with similar metrics. While not presenting a complete picture of what is effective, these implicit discussions identify aspects of effectiveness, often in the context of a specific example, such as the definition listed in Table 8.1 by Boklan and Janusz-Pawletta (2017).

Of the literature dataset, 20 articles explicitly discussed effectiveness in general and in the context of related terms to cooperation. Of these, only 5 articles have explicitly defined effective cooperation. Some of the articles were coded

Table 8.1 Sample of text coded as implicit discussions of effective cooperation

Implicit discussion of effective cooperation	*Source*
"While proving the hypothesis of this paper, that the EAEU could serve as a legal framework for more effective cooperation on transboundary waters in Central Asia, the answer given could only be positive, but only if taking into account several conditions, the EAEU should comply with to reach this goal. The first condition is to introduce the main principles and mechanisms of international water law into the existing EAEU legal framework. Second would be to strengthen the legal basis to enable cooperation between EAEU states on environmental issues in general and transboundary water relations in particular. The current framework for transboundary water cooperation within the EAEU, unfortunately, is rather limited and out of date."	(Boklan and Janusz-Pawletta 2017, 437)
"Although it is unrealistic to expect that watercourse States could be compelled to establish strong and all-encompassing joint bodies from scratch, it is of the upmost importance that States enter into some kind of institutional arrangements to begin to build the mutual trust needed for effective cooperation in a transboundary context."	(Belinskij 2015, 5400)
"The assumption was that effective cooperative sharing of a common resource at the local level depends on some degree of common language, understanding skills, confidence, and opportunities (Cook 1995) so that contributions to conversations and decisions can be collective and equitable."	(Mirumachi and van Wyk 2010, 31)
"Strong political will and commitment, negotiation, financial resources, and meaningful technical cooperation will lead to successful transboundary cooperation in water management."	(Rasul 2014, 32)
"Grey et al. (2009, 19) put it well when discussing interventions in effective transboundary water cooperation: 'process is almost as important as product, at least in the early days.'"	(Zeitoun 2013, 147)

for both implicit and explicit discussions of effectiveness. Table 8.2 lists all 5 explicit definitions of effective cooperation identified in the literature sample.

Unsurprisingly, there are very few explicit definitions of effective transboundary water cooperation and few implicit definitions in the literature sample. The explicit definitions of effective cooperation are more direct in defining the term and are more comprehensive in their discussions. Considering these definitions, we can begin to see themes including specific actions or outcomes of the cooperation, as well as measures drawn from international relations literature, such as the idea of problem solving and process effectiveness.

Comparatively, the 54 interviewees were directly asked to define effective transboundary water cooperation via the question: 'How would you define effective transboundary water cooperation? What do you mean by effective?' However, there is a wide range in the level of detail, comprehensiveness,

Table 8.2 All five explicit definitions of effective cooperation from the literature data sample

Explicit definitions of effective cooperation	*Source*
"UNDP (2010) argues that preventing inter-state water conflicts cannot be realized without effective cooperation in managing transboundary watercourses, built through strong and equitable structures and institutions for collaboration at national and regional levels. This cooperation should not be seen as a goal in and of itself, rather a required tool to meet the goals of the co-riparians, improve water governance and achieve clear progress towards security and sustainable development."	(Chikozho 2014, 74)
"Based on these different angles and elements associated with effective cooperation, we take the following working definition of effective water cooperation: a collaboration in which to or more parties find a negotiated compromise on maximizing mutual gains and achieving joint wins for all parties involved, resulting in the availability of an acceptable quantity and quality of water for health, livelihoods, ecosystems and production, coupled with an acceptable level of water-related risks to people, environments, and economies."	(Huntjens et al. 2016, 19)
"Effective cooperation on an international watercourse is any action or set of actions by riparian states that leads to enhanced management or development of the watercourse to their mutual satisfaction."	(Grey, Sadoff, and Connors 2009, 19)
"This paper offers a definition of effective cooperation that focuses on providing measurable water security benefits to individual basin states. It recognizes that effective cooperation is a multi-stage continuum of relations between basin states and that many tentative, preliminary steps, such as agreeing to discuss problems and to share data collection and distribution, are necessary to make progress towards improving water security throughout the basin. More formal steps may follow, such as agreeing water allocations and developing a legal management framework."	(Tarlock 2015, 9–10)
"Brown and Ashman reinforce this view, noting that 'Cooperation must span gaps in culture, power, resources, and perspective. … Effective cooperation requires some degree of mutual influence that allows that all the parties to influence that allows all the parties to influence and be influenced' (1996, 1467). … Here, the effective cooperation based on riparian compliance, goals, interests and problem-solving characteristics has been differentiated from the unchallenged typical indicators of cooperation such as the signing of a treaty or establishment of a river basin initiative."	(Zeitoun and Mirumachi 2008, 304 and 312)

and clarity of the discussion between the interviewees. Responses varied from straightforward and general definitions to aspects that would enable effective cooperation, as well as those that acknowledge the inherent variability in defining effective cooperation. For example, Dr. Robert Varady of the University of Arizona gives a more general definition: "I guess there would be several elements to that effectiveness. The first one is the obvious one. Are things better than they were before? It's [a cooperative process] that improves the situation, and continues to do that, and is able to retain the improvement. And, maybe even improve the improvement, make it even better. But, clearly, there's a strong temporal element to it" (Varady 2018).

This compares to the discussion with other interviewees, who argue that effective cooperation has several attributes, including scientifically informed decision making and political will, where political will is driven by the understanding of actor needs and that there is support for cooperation. These compare distinctly to the insights provided by Marguerite de Chaisemartin, of University of the Pacific, where she notes that the definition of effective cooperation depends on the result: "It depends on what the result sought is … the elements that make cooperation 'effective' would differ depending on what you want to achieve" (de Chaisemartin 2018). Just from these three examples, the challenge in defining effective cooperation is visible.

Overall, looking at the implicit and explicit definitions of effective cooperation only presents a narrow view of what effective cooperation is. More detail is needed to better understand how effective cooperation is generally perceived, what are the most important aspects that contribute to effective cooperation, and how we can then evaluate the degree of effectiveness for transboundary water cooperation. As with the data on cooperation (see Chapter 5), I coded the implicit and explicit discussions of effectiveness by their *aspects of effectiveness*. Through these, we can identify trends, similarities, and differences between the discussions and definitions of effectiveness, as well as between the data sources.

Aspects of effectiveness

The aspects of effectiveness are the word, phrase, or sentence(s) representing a standalone concept of effectiveness or effective cooperation. These are similar to the characteristics of effective cooperation that Huntjens et al. (2016) identify through their literature review when working toward defining effective cooperation for use in their multi-track diplomacy framework. By coding for aspects of effectiveness, we can determine what the literature, academics, and practitioners perceive as contributing to the effectiveness of cooperative arrangements. These can then inform a new model for evaluating effective cooperation based on these perceptions, developed in the next chapter. However, in coding these aspects, a challenge that I encountered was that effectiveness is subjective and context dependent. What is considered effective cooperation in one context could be considered ineffective cooperation in another river basin or by another evaluator. Therefore, this chapter aims to categorize the aspects into broad categories and look at the more commonly referenced aspects that could be more applicable or customizable to a particular context. It is important to note prior to explaining the results of this analysis that it does not determine or establish causality between the aspects and whether or not cooperation is effective.

Table 8.3 Top most commonly referenced aspects of effectiveness in the data broken down by implicit and explicit discussions and between interviews and literature data

Effectiveness implicit – literature only	*Effectiveness explicit – literature only*	*Question defining effective cooperation – interviews only*	*Effectiveness explicit – literature and interviews combined (by number of references)*
Institutional capacity	Achieve outcomes	Achieve outcomes	Achieve outcomes
Achieve outcomes	Institutional capacity	Participation	Participation
Define outcomes	Define outcomes	Define outcomes	Mutual satisfaction
Dispute resolution	Participation	Context	Multi-scale
Adaptability	Multi-scale	Information and data	Define outcomes

Within the contextual units of implicit and explicit definitions, we can see differences in the most commonly found aspects of effectiveness. Table 8.3 lists the top five most commonly referenced aspects of effectiveness in the literature, interviews, and implicit and explicit definitions.

Overall, the three most commonly coded aspects of effectiveness are: *achieve outcomes, institutional capacity*, and *participation*. Text coded as *achieve outcomes* states that effectiveness is achieving general goals, outcomes, benefits, or results, which are not defined. For example, Prasai and Surie (2015, 30) write of the Ganges basin that it is needed to "meaningfully engage and produce tangible results on water cooperation." Alternatively, the code also includes text where a specific goal, outcome, benefit, or result is defined. Maria Apostolova of the Global Environment Facility captured the sentiment of both aspects of the *achieve outcomes* node by stating in response to the question to define effective cooperation: "I think like a more qualitative [sense of effectiveness]. A positive impact on integrated water resources management, like optimizing the use and protection of the resource and avoiding conflict. But this is very difficult to measure. … It's more to accomplish outcomes that were set and agreed among the countries" (Apostolova 2018). In all divisions of the data, with the exception of the implicit discussions in the literature, achieving outcomes and results seems to be the most important aspect of effective cooperation. This is interesting when compared with the most commonly referenced key element of cooperation (see Chapter 5), which is *treaty or agreement*. This is a common indicator of cooperation. However, having a treaty or agreement is not within the top-listed aspects of effective cooperation. This demonstrates that while having an agreement increases the potential for cooperation, it needs to be implemented and its stated objectives achieved to be effective, so as not to be a 'paper tiger.'

Institutional capacity, as an aspect of effectiveness, is similar to the coding definition within the key elements of cooperation analysis. It includes text that specifically mentions 'institutional capacity' or the combination of agreements, river basin organizations, or institutions more generally. For example, Bernauer (2002, 14) writes, "Marty proposed that cooperative arrangements are likely to be more successful if they link international river commissions and national level authorities, thereby ensuring financial and political support within riparian country bureaucracies." Interestingly, institutional capacity is the most common aspect in implicit discussions – such as the Bernauer quote – and the second most common for explicit definitions in the literature. However, institutional capacity is not among the top referenced aspects of effectiveness in the interviews, nor in the combined data (see Table 8.3). I would argue that this difference is due to the fact that institutional capacity is a prominent component of cooperation in the literature and is more easily measurable than many of the top aspects of effectiveness from the interviews only. Although less prominent, we can see a similar diversion when comparing the top aspects of effectiveness between the academics and practitioners in the interviews alone (see Table 8.4). *Institutional capacity* is the 3rd most commonly referenced by academics but is mentioned 12th most commonly by the practitioners. While, as mentioned previously, we cannot draw causality from these data, I would argue that this difference is due to the fact that much of the academic perspective is drawn from the literature. Whereas, practitioners may draw their perspective from experience, which might illustrate that other aspects of effectiveness are more important for effective cooperation.

The last of the three most commonly referenced aspects of effectiveness overall is *participation*. Text included in the code for *participation* includes discussions of having all riparians involved in a cooperative arrangement, including non-state actors and other stakeholders, holding inclusive and participatory processes, as well as the active participation of actors in the process. For example, Rahaman

Table 8.4 Top aspects of effectiveness referenced in the interview data separated by the interviewee category

Top aspects of effectiveness by interview category				
Academics	*Practitioners*	*Academic/ Practitioner*	*Academic/NGO*	*Practitioner/NGO*
Achieve outcomes	Achieve outcomes	Achieve outcomes	Achieve outcomes	Achieve outcomes
Context	Participation	Context	Context	Power/Political will
Institutional capacity	Information and data	Participation	Decision making	Define outcomes
Participation	Regional pressure	Institutional capacity	Future needs	Participation
Dispute resolution	Implementation	Time	Consensus	Implementation

(2012, 489) states, "the basic foundation of effective transboundary water management depends on the participation of all riparian states in the management of the basin." We can see a different statement from Lesha Witmer of Women for Water Partnership, "I think the second part [of defining effectiveness] has to do with who do you involve. If you keep it very bureaucratic. Like it can only be the elected or appointed officials, but it's not really addressing issues of the people on the ground. Then, in the end, it's a nice process and a nice procedure, but it's not effective" (Witmer 2018). While this quote highlights that limited participation is ineffective, the converse is that with wider participation, there is more effective cooperation. Participation seems to be more heavily referenced in the interviews – more so by the practitioners than the academics – than in the literature; in fact, participation is not even within the top aspects for the implicit discussions of effective cooperation in the literature.

Aside from these three most commonly referenced aspects, we can see other similarities between the top aspects in the implicit and explicit discussions of effectiveness as well as between the literature and the interviews (see Table 8.3). A commonly mentioned aspect in both data sources is *define outcomes*; these data reference text that is calling for joint identification of problems and/or solutions. This aspect reflects the need for a standard to compare and evaluate effectiveness against, and it presents one option – the solutions defined by the parties can be evaluated by whether they have been achieved. Throughout both the literature and the interview data, the aspects reflect the potential for variability more so than the key elements identified for the definition of cooperation; these include *multi-scale*, *context*, and *adaptability*. *Multi-scale* includes text that references the need for basin-wide efforts, that cooperative processes should occur at the appropriate scale (often without reference to what that scale is), and that emphasizes multi-scalar governance as a part of cooperation. *Context* addresses ideas that effectiveness is relative and dependent on the situation's hydrology, politics, history, and so on. *Adaptability* predominantly captures text that recognizes the need for cooperative processes to be adaptable to future change and variability – the ability to adapt could contribute to its future stability.

However, there are also differences between the top aspects. Notable is the inclusion of *dispute resolution* in the implicit literature discussion and by the academics in the interviews. Here, *dispute resolution* includes text that states that there is a mechanism in a treaty or institution for dispute resolution. This should not be confused with the specific outcome of reduced or resolved conflict, which is coded separately. Practitioners addressed *information and data*, *regional pressure*, and *implementation* in their discussions of effective cooperation. Data and information included joint data collection, sharing data, agreeing and accepting the validity of shared data, as well as mechanisms for data and information exchange. Interestingly, many of the practitioners cited the role that the regional environment has played in creating effective cooperation, such as the Water Framework Directive, which requires joint basin planning and basin-wide efforts, which is coded as *regional pressure*. Lastly, *implementation* captures the idea that cooperative processes are making progress toward resolving their problems or are working toward their goals but have not yet achieved them; often, this idea was expressed as: states are working to implement what was agreed upon in a treaty. If we look at the differences between the aspects through the lens of the Four Frames of Cooperation, we can see that the aspects of effectiveness mentioned in the literature and by the academics fit within the institutional (*institutional capacity*) and legal frames (*dispute resolution*), whereas the practitioners include aspects related to the greater context (*regional pressure*) and outcome (*implementation*). *Information and data* tend to fall between the institutional and relational frames, as it encompasses the mechanisms and actions to collect and share data, but also the trust and acceptance of data provided by other parties.

The final point I would like to discuss regarding the data on aspects of effective cooperation is with respect to whether there is a difference between groundwater and surface water. Thus far, this book has addressed transboundary water cooperation and its definition as broad enough to be adaptable between surface water, groundwater, and situations of conjunctive management. Data on groundwater are far less prevalent; most articles and grey literature used in the content analysis only address surface water. Therefore, in the interviews, the experts on groundwater were asked an additional question: whether the definition of effectiveness they provided was applicable to both surface and groundwater or whether there would be a difference in evaluating the effectiveness of cooperation on surface water versus groundwater. The top five aspects most commonly referenced by the experts on groundwater are *achieve outcomes*, *shared model*, *implementation*, *multi-scale*, and *participation*. For the most part, these are the same top aspects coded in general in the data. The key difference is the inclusion of the *shared model* as the second most referenced aspect. Nearly all the experts on groundwater – as well as other academics and practitioners – highlighted that often, very little is known about the sub-surface, including how surface waters and groundwaters are hydrologically connected. Having a shared conceptual model that is agreed upon by the riparians is vital for effective cooperation over groundwater, as it is difficult to cooperate and agree on solutions without a better understanding of the resource. However, one interviewee countered this view and proposed that even without as much data as is available for surface water, actors should still be able to cooperate and co-govern an aquifer. The lack of data should not be used as an excuse to not cooperate over groundwater. In a way, perhaps the lack of data limits states' ability to politicize or securitize groundwater, as can be done with many surface waters. This could open space for effective cooperation or, conversely, the uncertainty could add risk, leading to less willingness to cooperate effectively. Overall, almost all interviewees who were asked about groundwater believed that a definition of effective cooperation would be the same for both surface- and groundwater; however, when considering the objectives and details that are context specific, the difference between the two resources would become apparent. Maria Apostolova of the Global Environment Facility succinctly stated: "I think in a broader context they should be pretty much the same. And when you go down to measurements and more specific indicators, you differentiate between the two resources" (Apostolova 2018).

There is no definitive way to determine, at this point, whether these aspects actually *cause* effective transboundary water cooperation. This chapter draws from the literature and practical experience of academics and practitioners to identify what is perceived as the most common aspects of effective cooperation in order to develop an understanding of what effective cooperation looks like for transboundary waters. As with the key elements of cooperation, I hypothesize that the most commonly referenced aspects align with the literature and with the general understanding of effectiveness, that is, to achieve an outcome or result. The aspects discussed here are not the only aspects of effectiveness – for the full list of aspects identified in the data, see Appendix B.

Examples of effective cooperation

In addition to asking experts to define effective cooperation, I was also interested in learning what cooperative processes they thought were effective to see if there were any similarities between the basins or aquifers. The experts were asked: Do you have an example of a basin that exemplifies effective transboundary water cooperation for you? Why this basin? Table 8.5 lists all the basins and aquifers identified as having effective cooperation or at least aspects of effective cooperation.

Generally, the most frequently referenced basins and aquifers were similar between the academics and the practitioners. The basins and aquifers most frequently referenced as effective were the Geneva Aquifer, Columbia River, Danube River, Senegal River, and the Rhine River. The academics tended to reference the Columbia most, whereas the practitioners mentioned the Senegal River. While there is no means to determine why this minor difference exists, it is likely due to the personal experience of the individuals I spoke with. I find this difference interesting given that the Columbia River has no river basin organization and little in the way of shared institutions. The Senegal River, on the other hand, has a highly integrated management structure with a strong

Table 8.5 List of the basins and aquifers mentioned as having effective cooperation or aspects of effective cooperation by multiple interviewees

Basins and aquifers with effective cooperation or aspects of effective cooperation[a]		
Al Disi Aquifer	Great Lakes Basin	Rhine River
Colorado River	Guarani Aquifer	Senegal River
Columbia River	Indus River	Stampriet Aquifer
Danube River and ICDPR	Mekong River	
Geneva Aquifer	Nubian Sandstone Aquifer	

[a] This is based on the data collected and is not an evaluation made by the author nor the method of evaluation of effective cooperation as developed in this book.

river basin organization – the Senegal River Basin Development Organization (OMVS) – and has jointly owned and managed infrastructure. These two basins have very different cooperative processes, yet both are identified as effective basins.

When asked why they believed the Columbia was an example of effective cooperation, interviewees often cited that the Columbia is frequently used in the literature as an example of good transboundary water cooperation. This is because the treaty, signed in 1964, employs benefit sharing to distribute equally hydropower and flood control benefits. The fact that First Nations and Native Americans have been historically left out of the agreement, as have environmental issues, reduces the perception of effectiveness today. However, even with these missing aspects, the long-term relationship between the USA and Canada is seen as having trust and good dialogue. The states have relatively equal bargaining power and capacity, for example, scientific knowledge, negotiation, and so on. We can compare these reasons for why Columbia is considered to have effective cooperation with the responses about the Senegal River. Also mentioned extensively in the literature, the Senegal River is jointly managed between Guinea, Mali, Senegal, and Mauritania through the OMVS, which was established in 1972. The OMVS implements equitable sharing of the benefits and risks between the states, which jointly own the major infrastructure along the river. The basin-wide management has an integrated scope and includes environmental protection and concern for local populations in addition to infrastructure management and water allocation. Ultimately, interviewees cited that there is apparent trust between the states, which have relatively equal capacity. There is also trust in the OMVS and the process of cooperation and management; this includes trust in the data collected to make management decisions.

Looking at these two brief examples of effective cooperation – the Columbia and Senegal Rivers – we can identify similarities in the cooperation, despite them having distinctly different legal agreements, institutions, and contexts. It is apparent that effective cooperation is perceived to be cooperation with strong positive relational and outcome frames. Trust, relationships, and capacity over time support and demonstrate effective cooperation, as does the ability of the two cooperative processes to share benefits and execute the outcomes defined, even though these outcomes are different between the basins. Similar comments regarding the relational and outcome-framed aspects of cooperation were made with respect to several of the other basins listed. However, one variation was responses on whether there was effective cooperation in the Indus River Basin. Interviewees generally expressed that cooperation was effective because it had reduced conflict in the basin, especially when considering the tense, sometimes hostile, relationship between India and Pakistan.

Looking at these data about examples of effective cooperation overall, the most commonly referenced aspects are: *achieve outcomes*, *time*, *institutional capacity*, *process*, *financial support*, and *context*. In general, these are very similar to the top aspects referenced in answers to the defining effectiveness question. Given the subjectivity of effectiveness, it is logical that time and context are prominent. The new aspects not previously defined are *process* and *financial support*. *Process* is coded by the same definition used for the key element of cooperation. However, in this case, it is used in the context of effective cooperation, where process is decision making, problem identification, and problem solving when discussed in a general or non-specific manner that is often used to embody the relationships required to initiate, develop, implement, and maintain cooperation. *Financial support* includes discussions about how adequate and sustainable financing is needed for impactful cooperation. Overall, I would argue that through these data, effective cooperation is not just the normative procedural elements that are often easily measured, but rather a definition of effective cooperation also needs to include relational and outcome aspects of effectiveness.

Defining the lack of effectiveness

As we have seen, defining effective transboundary water cooperation is a difficult task. Rather than define what effective cooperation is, authors often define what effective cooperation is not. Within the data, text often defined ineffectiveness or components of ineffective cooperation. It is worth looking briefly at several examples to try and learn what effective cooperation is not. The following is a select list of quotes from the data on the lack of effectiveness:

- "[i]f monitoring and data sharing do not feed into the decision-making process, confidence in cooperative processes may actually be eroded due to the lack of tangible benefits" (Chenoweth and Feitelson 2001).
- "A number of obstacles exist in developing an effective agreement to share water resources between governments. These obstacles include conflicts in internal water laws and politics of the parties, an incomplete knowledge of the water resource availability and demand within the shared basin, and conflicts between the internal economic policies of the parties" (Environmental and Water Resources Institute 2012).
- "Much of the literature either explains why cooperation has not succeeded or details examples of 'paper' coordination. That is the signing of agreements among basin states, but with little attention given to implementation" (Tarlock 2015).
- "[i]f there is no trust, no capacity, no data, no investment, and engagement or commitment in time, then you don't have effective cooperation" (Salamé 2018).
- "[b]ecause they have to keep negotiating for every step they take. It's not like a normal management plan for two or three years that everybody's contributing to. No, they have to negotiate every step of the way. That's not effective. It looks nice from the outside, but it's not effective" (Witmer 2018).

These examples of ineffectiveness parallel the data on effectiveness. The first quote by Chenoweth and Feitelson (2001) talks about the 'lack of tangible benefits' or outcomes, as well as the need for process and data and information sharing. The second quote from the Environmental and Water Resources Institute (2012) acknowledges that the role of context – such as domestic laws and politics, existing conflicts, and economic policy – will affect whether an agreement is effective. Tarlock (2015) points out a critique of many indicators of cooperation that focus on the signing of an agreement and the mechanisms it contains; while important, without implementation, they cannot be effective. The first interviewee notes that without the relational elements, data and information, time, and financing, there is no effective cooperation that mirrors the aspects of effectiveness visible through the examples discussed above. Lastly, the final quote highlights the need to have an effective process; a cooperative process that requires negotiation regularly is an ineffective process and likely one that lacks trust and strong relationships between the parties. By considering what is ineffective or a barrier to effective cooperation, this chapter aims to strengthen the understanding of what effectiveness means in order to better construct a model for evaluating effective cooperation.

Barriers to effective cooperation

We can learn more about what constitutes effective cooperation by further exploring what effective cooperation is not, through what are perceived to be the most common barriers to effective cooperation. Experts were posed the question: What are the common barriers you see in basins or aquifers preventing effective transboundary water cooperation? The responses to these questions were coded according to the specific themes common throughout the data. Overall, 29 distinct barriers were identified in 45 responses. There is no space here to discuss all 29 barriers identified; therefore, the complete list with descriptions is included in Appendix C. The top five most commonly referenced barriers, broken down between the interview categories, are included in Table 8.6.

Table 8.6 The top most referenced barriers to effective cooperation broken down by the interview category

Commonly referenced barriers to effective cooperation by interview category				
Academics	*Practitioners*	*Academic/Practitioner*	*Academic/NGO*	*Practitioner/NGO*
Context	Context	Context	Context	National level
Political will	Financing	National level	Political will	Financing
Stakeholders	Political will	Lack of knowledge	Capacity	Communication
National level	Competing perceptions	Competing perceptions	National level	Political will
Lack of knowledge	Trust	Communication	Lack of knowledge	Lack of knowledge

The top commonly referenced barriers across all the data are *context*, *political will*, *national level*, *financing*, and *lack of knowledge*. There are only a few differences between the various categories of interviewees. *Context*, unsurprisingly, was the most commonly referenced barrier to effective cooperation. As with the aspects of effectiveness, *context* is broadly defined to include the political situation, historical context, climate, culture, economy, and so on. A little over 50% of the respondents mentioned context or a contextual factor as a reason why effective cooperation is difficult. Dr. Robert Varady, of the University of Arizona, sums up nicely one impact that context can have in encouraging or increasing the difficulty of cooperation: "Round up the usual suspects, like the differences and administrative procedures, legal procedures, cultural practices, farming practices, state of development, GDP per capita, all those things. The more pronounced the difference, the greater the impediment to coming to reasonable agreements. If you have two countries bordering each other with very different rates or measures of all those things that I mentioned … that's going to complicate the thing. As opposed to two very similar countries, where people speak the same language, and the landscape is similar, and the economic status is similar, the GDP per capita, the population growth rate. If all those things are similar, there's a better chance of coming to an accommodation" (Varady 2018).

Political will differs from political context, as it addresses the political intent, willingness, pressure, and so on to engage or not to engage in a cooperative process over transboundary water. Most of the interviewees with responses coded for *political will* explicitly state the term. Many interviewees stated that one of the most significant barriers is having (or not having) the political will. For example, one interviewee who leads a large river basin organization noted that political commitment or political will is a regularly discussed idea in the field. Political will can both encourage and enable effective cooperation and has the potential to impede effective cooperation if the political commitment is inadequate or lacking. From the opposite perspective, political commitment can be nurtured and developed through cooperation. Political will and commitment are intertwined with cooperation and the lack of effective cooperation, with both elements having the ability to impact each other. Neither of these two barriers – *context* and *political will* – is likely to be a surprise, as these are two frequently discussed factors that drive or inhibit cooperation in the literature.

National level was an interesting theme that came from the data, and it encompasses a few ideas related to national level governance and national positions. First is the idea that national positions and securitization of water resources at a national level inhibit the willingness to cooperate. Countries are focusing on national positions without actually negotiating. Second, there may be cooperation between states; however, the implementation is hampered by inadequate national laws and policy to implement the decisions made through cooperation. Lastly, interviewees discussed the opposite idea that cooperation was hampered by the involvement of a larger scale – often national or regional – in problems that could be effectively solved by local or smaller scale efforts. What we can draw from this complex and nuanced theme is that cooperation should be at the appropriate scale for the system and problem at hand, national policies need to support international cooperative efforts as well as local cooperative efforts, and a shift in thinking is needed from positions and the idea of water as a matter of national security to interests and needs that will allow for actors to better identify shared benefits.

The last two most commonly referenced barriers are *financing* and *lack of knowledge*. *Lack of knowledge* refers to the lack of reliable data and limited scientific understanding, in addition to the lack of data collection. *Lack of knowledge* was frequently mentioned in reference to groundwater, given that there is very little data globally on groundwater and aquifer systems. *Financing* was another nuanced barrier that was discussed in the data. Like *national level*, there are a few ideas within this theme. The straightforward response was that there needs to be adequate, sustainable, and predictable funding for cooperation to be effective. However, this can contribute to an over reliance on external funders, which in and of itself is a barrier, as this can lead to a donor priority-driven process rather than a stakeholder-driven process. In addition, external money may be the contributing factor as to why the cooperative process is occurring. This could act as both a driver and a barrier to cooperation. External funding can encourage the development of cooperation, but if trust, relationships, and national desire do not develop, then the process is not truly effective, as it is strictly being driven by external funding. Without this funding, the question is: will this process be maintained? Lastly, it was mentioned that perhaps having too much money is a barrier to effective cooperation. The example given was the Mekong River Commission, where there is significant funding provided; however, the funding emphasizes conducting as many studies as possible within the basin to meet donor expectations, yet many of these studies are not properly implemented.

There are a few other barriers mentioned between the different interview categories, as shown in Table 8.6. However, there does not seem to be a trend in the types of barriers identified between academics and practitioners, as there are with the key elements of cooperation and the aspects of effectiveness. The remaining most referenced barriers mentioned, but not discussed in detail here, include *trust*, *communication*, *stakeholders* (which refers to not having all riparian and stakeholders involved in the process), *competing perceptions*, and *capacity* (which refers to the ability of actors to participate in the process, in negotiations, in data collection, etc.). In general, we see that most of the barriers referenced with respect to effective cooperation deal with the relational and institutional frames of cooperation, as well as the greater context and scale of the cooperative process. Ultimately, with such a complex topic, it is no surprise that the barriers are just as deeply nuanced, equally complex, and context specific as effective cooperation itself.

Examples of basins and aquifers with barriers to effective cooperation

To expand on what is perceived as barriers to effective cooperation, we can consider similarities and differences in cooperative processes that are considered

Table 8.7 The list of basins and aquifers listed as having barriers to effective cooperation based on multiple responses from interview data

Basins and aquifers with barriers to effective cooperation based on multiple responses[a]		
Aral Sea Basin	Helmand River Basin	Tigris-Euphrates River Basin
Ganges-Brahmaputra-Meghna River Basin	Mekong River Basin	US-Mexico Groundwater
Guarani Aquifer	Nile River Basin	

[a] This is based on the data collected and is not an evaluation made by the author nor the method of evaluation of effective cooperation as developed in this book.

ineffective or lacking aspects of effective cooperation. Experts were asked: Do you have any examples of basins that do not have effective transboundary water cooperation? Table 8.7 lists all the basins and aquifers identified as having barriers to effective cooperation by multiple respondents.

Overall, the Aral Sea and the Nile River were the two most commonly referenced examples of basins with barriers to effective cooperation. The others include the Mekong, the Ganges-Brahmaputra-Meghna, and the Guarani Aquifer, although these were referenced moderately less than the Aral Sea Basin and the Nile River Basin.

The experts gave similar reasoning for why the cooperative processes in the Aral Sea Basin and the Nile River Basin had barriers to effective cooperation. When asked why the Aral Sea's cooperative process was not functioning effectively, the interviewees' responses were centered on a few key points. First, nearly all respondents pointed out that there is cooperation in the basin; however, it is not effective cooperation. The process is considered a failure since there has not been much in the way of positive outcomes – particularly environmental improvements in the basin, for example, the decline in the Aral Sea level, decline in flow in the Amu and Syr Darya Rivers, and so on. Key issues spurring cooperation have not been solved, such as conflicts over hydropower development and irrigation. This failure is argued as being due to a focus on national positions, the tendency to blame other stakeholders for issues with the water resources, and the overuse of water in general. The lack of reliable data and trust in the available data contributes to these factors. Ultimately, most respondents stated that there is a lack of trust, communication, and capacity in the basin. States mistrust each other and the institutions in place; therefore, to increase the effectiveness, there needs to be more confidence in the institutions and the process by the parties involved. The states have differing levels of development, which impact their respective capacities to negotiate, collect data, and understand the potential shared benefits; this can inhibit the ability to develop trust between the parties.

The Nile River Basin has a long history of sharing transboundary waters with varying levels of cooperation. When asked why the Nile was not considered to have effective cooperation or what aspects inhibited effective cooperation, the interviewees acknowledged that there is cooperation and there has been significant progress in the basin over the last several decades; however, the cooperation is not effective in the basin as a whole. The history of colonization and the historical agreements splitting the waters of the Nile between Egypt and Sudan have had a lasting impact on the ability of current processes to be effective. This history has led to uneven development in the basin, creating power differentials, which has led to hegemonic development and use of the Nile waters. Similar to the Aral Sea, the recent discussions on water security illustrate a focus on national positions, rather than identifying shared benefits. Ultimately, most of the respondents highlighted that not all countries are fully participating in the process; even those that are participating may be doing so hesitantly. There is a lack of trust in the basin, particularly between the upstream and downstream states. This lack of trust in the downstream states has encouraged unilateral development in the upstream states; consequently, this increases the mistrust of the downstream states. The lack of trust and full participation in the process occurs despite active dialogue between the countries and ample reliable data. Despite all these comments on the ineffective aspects of cooperation in the Nile, one interviewee noted that we must consider the cooperative process over time. In the almost 100 years of formal agreements and 'cooperation' in the basin, there has been significant progress. Cooperation is a process that ebbs and flows; therefore, when evaluating effectiveness, the timeframe must be considered.

Looking at the two brief examples of ineffective cooperation – the Aral Sea and Nile River basins – we can identify similarities in why the cooperative processes are not considered effective. What is apparent is that cooperation is perceived to be ineffective because both basins have weak and sometimes negative relational elements of cooperation. Trust, communication, and fully engaged participation are lacking in the basins. In the Aral Sea Basin, institutional elements also need to be strengthened with an increase in states' capacities and in data collection and sharing. A few individuals also highlighted the legal element in the Nile, arguing the cooperation would be more effective once the *Cooperative Framework Agreement* is enforced. Similar comments regarding the relational framed elements of cooperation were made with respect to several of the other basins and aquifers listed.

It is also interesting to note that multiple interviewees mentioned the Mekong River Basin as having both effective aspects of cooperation and as having barriers to cooperation. It is considered effective in that there is political will from the lower basin states, which is backed by funding from each country in addition to donor funding. Furthermore, the cooperative process has been maintained for a long history in the lower basin. However, the cooperative process is seen as being structured in a not particularly effective manner, which contributes to the limited outcome effectiveness. A further barrier is a fact that Myanmar and China are only at observer status in the Mekong River Commission and do not fully participate. China has recently established, in 2015, an alternative process in the Lancang-Mekong Cooperation forum, which has been addressed with mixed perspectives as to its contribution toward effective cooperation in the basin. This mixed example of both effective aspects and barriers, I believe, is representative of most shared waters, where there is likely to be both aspects of effective cooperation and barriers working in the basin.

This discussion of why interviewees perceived specific basins to be ineffective in their cooperative process highlights discrepancies between perception and what is measured in definitions of transboundary water cooperation. In light of this, are the measures of cooperation being used truly capturing what is perceived about cooperative processes? For example, let's compare the above discussion on the Nile River Basin's barriers to effectiveness to the results of methods for measuring operational arrangements for cooperation presented in Chapter 3. For Method 1, which uses the methodology for SDG Indicator 6.5.2, Uganda has a 90% coverage of operational arrangements for cooperation because the Nile basin country unit (BCU) is calculated as meeting all the criteria for operational cooperation. For Method 2, the Nile BCU in Uganda is given an Operational Level 5 for meeting the same criteria as in Method 1. Lastly, Method 3 categorizes the cooperation in the Nile BCU of Uganda as 77% in Emerging Comprehensive Cooperation and the sub-basin of Lake Victoria as 21% in Continuing Comprehensive Cooperation. These analyses present the cooperative process in the Nile basin – at least within Uganda – as operational and progressing along the cooperation continuum. One could easily consider these results and assume that cooperation in the basin is effective. Yet, this is vastly different from the perspective provided by the interview data that the cooperation in the Nile, while progressing, is currently not effective.

This section highlights what are perceived to be the reasons why cooperation may not be considered effective. Furthermore, it illustrates the challenge of measuring cooperation without having a clear understanding of what is being measured and the goals of the methodology. Various definitions of cooperation lead to multiple, often competing perceptions of what is considered 'good' or 'bad' – or more simply, what is effective.

Bringing it all together

The Four Frames of Cooperation (see Chapter 6) defines transboundary water cooperation as the interactions between actors over shared waters that result in establishing mutually beneficial outcomes through a decision-making process; this process could include formal and informal legal and institutional mechanisms, depending on the scale and context. Cooperation can be explained through this definition by identifying elements within the four frames at the appropriate scale and context. The four frames are legal, institutional, relational, and outcome. The implicit and explicit discussions of effectiveness were coded based on the frame(s) of cooperation that they addressed in order to determine what frames of cooperation were perceived to be the most important aspects of

Table 8.8 Implicit and explicit discussions of effectiveness in the literature and interview data with the number of references to the Four Frames of Cooperation

Number of references in implicit and explicit discussions of effectiveness in the interviews and literature		
Frames of Cooperation	*Implicit*	*Explicit*
Relational	54	54
Outcome	50	50
Institutional	32	24
Legal	24	19

effectiveness. Table 8.8 lists the total number of references to each frame made in both the literature and the interview data. This table illustrates the same conclusion we can draw anecdotally from the discussions on the examples of effective and ineffective cooperation: *in defining and evaluating the effectiveness of transboundary water cooperation, the relational and outcome aspects are perceived as being more relevant.*

This trend is consistent when the data is broken down further – into interviews and literature only; however, there is only a difference of two to five references between the relational and outcome frames with the legal and institutional frames in the literature dataset. This makes sense, given what we know from the literature review and the previous discussions. The legal and institutional aspects are generally more measurable and, therefore, are often used as proxies or indicators for transboundary water cooperation. This might also be the reason for the slight difference when comparing academics and practitioners. Both academics and practitioners mentioned outcome and relational aspects more frequently than the institutional or legal frames. The academics mention outcomes slightly more frequently than relational aspects, while practitioners mentioned relational aspects slightly more than outcomes. These differences are minimal and could simply reflect the individuals I spoke with; however, I found this slight difference to be interesting. These findings echo the discussions of effective regimes from international relations theory. Recall Figure 7.1, where Underdal (2002) demonstrates the evolution of regime effectiveness, which begins with *outputs* (regime formation) that lead to *outcomes*[3] (implementation and changes in behavior) and then to *impacts* (responses to changed behavior). Using the Four Frames of Cooperation, this model of evolving effectiveness can be adapted into Figure 8.1.

While this representation is clean and idealistic, the reality is anything but clean and idealistic. Effective cooperation does not progress in this manner, but these stages overlap, as do the Four Frames of Cooperation, which can operate simultaneously. Mitchell and Zawahri (2015) sum up the idea of complex reality with respect to effectiveness by arguing that international rivers represent a dilemma as they cannot be 'solved.' Rather, states are confronted with complex

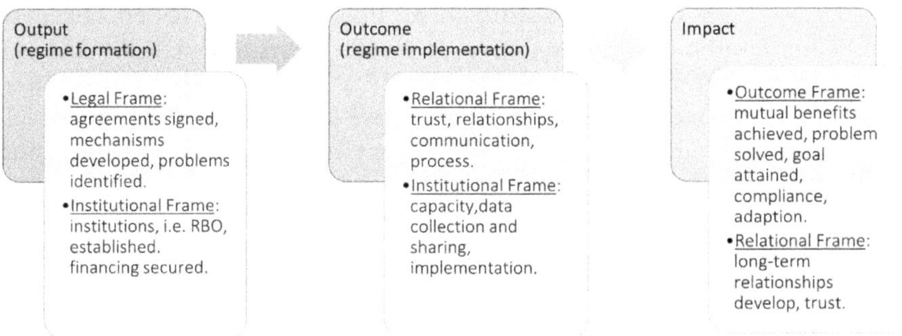

Figure 8.1 An adaption of the stages of regime effectiveness by Underdal (2002), aligned with the Four Frames of Cooperation.

Source: The figure is adapted from Underdal 2002 in Miles, Edward L., Steinar Andresen, Elaine M. Carlin, Jon Birger Skjærseth, Arild Underdal, and Jorgen Wettestad., *Environmental Regime Effectiveness*, p. 7, © 2001 Massachusetts Institute of Technology, by permission of The MIT Press.

relationships that must be managed because they lack a solution, or the solution is prohibitively expensive. Therefore, an institution is effective when it is able to assist states in managing a problem that involves continuous conflicts of interest. It is the process that enables effective cooperation to define and reach the outcomes while continuously adapting and redefining the problems, outcomes, and even the process. Ultimately, the effectiveness of all Four Frames of Cooperation needs to be incorporated in the model for evaluating effective cooperation, with an emphasis placed on the aspects of effectiveness within the relational and outcome frames.

Thus far in Part III, we have explored the literature on effectiveness and effective cooperation (see Chapter 7). In this chapter, we have taken a comprehensive look at perceptions of effectiveness through the content analysis of the sample of literature and interview data. What we can draw from this is that defining effective cooperation is complex. The term is context dependent, as well as subjective. So, how is effective transboundary water cooperation defined? Definitions vary considerably; however, the central aspects revolve around the idea that effective cooperation *achieves equitable results or outcomes through a cooperative decision-making process that has trusting relationships with stakeholder participation and communication. This process is supported by the appropriate legal and institutional arrangements, where the outcomes can be the resolution of the initial problems, reaching the goals of the parties, or complying with mutually agreed-upon principles in a legal arrangement.*

This definition is not operational, as it does not characterize a baseline or a standard to measure whether it is effective or not in achieving the outcomes, nor whether the process was effective. Rather it summarizes the aspects of effective cooperation. The investigation of the most commonly referenced aspects of effectiveness as well as how these fit within the Four Frames of Cooperation begins to develop a structure for understanding what effectiveness looks like in transboundary water cooperation. However, we are still unable to fully answer the question posed by Part III: What makes transboundary water cooperation effective. In addition, we still have not answered the questions posed by Underdal (1992, 228–29) presented in Chapter 7 as the core questions in defining and evaluating effectiveness:

> What precisely constitutes *the object* to be evaluated? (ii) Against which *standard* is this object to be evaluated? (iii) *How* do we operationally go about comparing the object to our standard; in other words, what kind of measurement operations do we have to perform in order to attribute a certain *score* of effectiveness to a certain object (regime)?

The aspects of effectiveness and the Four Frames of Cooperation represent the object to be evaluated, yet the standard to compare these aspects is still missing. The answers to these questions and the model for evaluating effective transboundary water cooperation are presented in the following chapter.

Notes

1 Please keep in mind when reading that this chapter focuses on defining *transboundary water cooperation,* not cooperation in the broader sense of the word. I use *cooperation* without the *transboundary water* modifier; however, I am referring to *transboundary water cooperation*. I will note when I refer to cooperation unlinked to transboundary water.
2 Expert interviewees included 22 Academics, 23 Practitioners, 3 Academic/Practitioners, and 6 Practitioners and Academics affiliated with NGOs: equaling a total of 54 interviews from 27 countries.
3 This outcome is defined as the changes in human behavior as a result of the regime. This differs from the definition of outcomes and the outcome frame that is used within the Four Frames of Cooperation, which are the benefits, goals, actions, or specific outputs of cooperation.

References

Apostolova, Maria. 2018. Research Interview.

Belinskij, Antti. 2015. "Water-Energy-Food Nexus within the Framework of International Water Law." *Water* 7 (10): 5396–5415. https://doi.org/10.3390/w7105396.

Bernauer, Thomas. 2002. "Explaining Success and Failure in International River Management." *Aquatic Sciences* 64 (1): 1–19. https://doi.org/10.1007/s00027-002-8050-4.

Boklan, Daria S., and Barbara Janusz-Pawletta. 2017. "Legal Challenges to the Management of Transboundary Watercourses in Central Asia under the Conditions of Eurasian Economic Integration." *Environmental Earth Sciences* 76 (12). https://doi.org/10.1007/s12665-017-6741-3.

Brown, L. David, and Darcy Ashman. 1996. "Participation, Social Capital, and Intersectoral Problem Solving: African and Asian Cases." *World Development* 24 (9): 1467–79. https://doi.org/10.1016/0305-750X(96)00053-8.

Chaisemartin, Marguerite de. 2018. Research Interview.

Chenoweth, Jonathan L., and Eran Feitelson. 2001. "Analysis of Factors Influencing Data and Information Exchange in International River Basins: Can Such Exchanges Be Used to Build Confidence in Cooperative Management?" *Water International* 26 (4): 499–512. https://doi.org/10.1080/02508060108686951.

Chikozho, Claudious. 2014. "Pathways for Building Capacity and Ensuring Effective Transboundary Water Resources Management in Africa: Revisiting the Key Issues, Opportunities and Challenges." *Physics and Chemistry of the Earth, Parts A/B/C*, Transboundary Water Cooperation: Building Partnerships (Part 2), 76–78: 72–82. https://doi.org/10.1016/j.pce.2014.11.004.

Cook, J. 1995. "Empowering People for Sustainable Development." In *Managing Sustainable Development in South Africa*, edited by Patrick FitzGerald, Anne Mc Lennan, and Barry Munslow. Cape Town [South Africa]; New York: Oxford University Press.

Environmental and Water Resources Institute. 2012. *Guidelines for Development of Effective Water Sharing Agreements*. Reston, Virginia: American Society of Civil Engineers.

Grey, David, Claudia Sadoff, and Genevieve Connors. 2009. "Effective Cooperation on Transboundary Waters: A Practical Perspective." In *Getting Transboundary Water Right Theory and Practice for Effective Cooperation*, 15–20. Stockholm: SIWI.

Huntjens, Patrick, Yumiko Yasuda, Ashok Swain, Rens de Man, Bjørn-Oliver Magsig, and Shafiqul Islam. 2016. "The Multi-Track Water Diplomacy Framework: A Legel and Political Economy Analysis for Advancing Cooperation over Shared Waters." Hague: The Hague Institute for Global Justice. www.thehagueinstituteforglobaljustice.org/thinking-doing/publication/the-multi-track-water-diplomacy-framework-a-legal-and-political-economy-analysis-for-advancing-cooperation-over-shared-waters/.

Mirumachi, Naho, and Ernita van Wyk. 2010. "Cooperation at Different Scales: Challenges for Local and International Water Resource Governance in South Africa." *The Geographical Journal* 176 (1): 25–38.

Mitchell, Sara McLaughlin, and Neda A. Zawahri. 2015. "The Effectiveness of Treaty Design in Addressing Water Disputes." *Journal of Peace Research* 52 (2): 187–200. https://doi.org/10.1177/0022343314559623.

Prasai, Sagar, and Mandakini Devasher Surie. 2015. "Water and Climate Data in the Ganges Basin: Assessing Access to Information Regimes and Implications for Cooperation on Transboundary Rivers." *Water Alternatives* 8 (2): 16.

Rahaman, Muhammad Mizanur. 2012. "Principles of Transboundary Water Resources Management and Water-Related Agreements in Central Asia: An Analysis." *International Journal of Water Resources Development* 28 (3): 475–91. https://doi.org/10.1080/07900627.2012.684311.

Rasul, Golam. 2014. "Why Eastern Himalayan Countries Should Cooperate in Transboundary Water Resource Management." *Water Policy* 16 (1): 19–38. https://doi.org/10.2166/wp.2013.190.

Salamé, Léna. 2018. Research Interview Skype.

Tarlock, Dan. 2015. "Promoting Effective Water Management Cooperation Among Riparian Nations." *TEC Background Papers*, no. 21. www.gwp.org/Global/ToolBox/Publications/Background%20papers/GWP_TEC21_web.pdf.

Underdal, Arild. 1992. "The Concept of Regime 'Effectiveness.'" *Cooperation and Conflict* 27 (3): 227–40.
———. 2002. "One Question, Two Answers." In *Environmental Regime Effectiveness: Confronting Theory with Evidence*, edited by E. Miles, A. Underdal, Steinar Andresen, Jørgen Wettestad, E Skjaerseth, and E Carlin, 3–43. Cambridge, MA: MIT Press.
UNDP, 2010. *Guidelines on the Joint Management of Transboundary Water Resources in the Arab World*. New York: UNDP.
Varady, Robert G. 2018. Research Interview.
Witmer, Lesha. 2018. Research Interview.
Zeitoun, Mark. 2013. "Global Environmental Justice and International Transboundary Waters: An Initial Exploration." *Geographical Journal* 179 (2): 141–49. https://doi.org/10.1111/j.1475-4959.2012.00487.x.
Zeitoun, Mark, and Naho Mirumachi. 2008. "Transboundary Water Interaction I: Reconsidering Conflict and Cooperation." *International Environmental Agreements: Politics, Law and Economics* 8 (4): 297–316. https://doi.org/10.1007/s10784-008-9083-5.

9 Evaluating effective transboundary water cooperation

A model for effectiveness

This is the final chapter of Part III. While Part II addresses a separate challenge, it is a foundation for the synthesis and answers to the questions posed in Part III – the main aim of this book. Ultimately, what makes transboundary water cooperation effective, and how can we evaluate its effectiveness? Before we dive into my answer to these questions, I want to summarize the main points of the preceding chapters, as these are the foundation and reasoning behind the model presented in this chapter. Part II with Chapters 4 and 5 considers two perspectives on how to define transboundary water cooperation through a literature review and through a qualitative content analysis of expert interviews ($n = 54$) and a sample of literature ($n = 273$). Using these data, I present a general definition of transboundary water cooperation:[1] *Interactions between actors over shared waters that result in establishing mutually beneficial outcomes through a decision-making process; this process could include formal and informal legal and institutional mechanisms depending on the scale and context.*

However, this is a generic definition and does not entirely capture the complexities of cooperative processes occurring over transboundary waters. Chapter 6 in Part II argues that we can develop place-specific explanatory neutral definitions through the Four Frames of Cooperation Framework, which provides a neutral structure for identifying cooperative elements within the Four Frames of Cooperation – legal, institutional, relational, and outcome. Part III begins with Chapters 7 and 8, which steers the conversation from explaining cooperation to evaluating cooperation. They argue that cooperation for cooperation's sake is not always 'good,' and that we need cooperation to be effective. However, there is no agreed-upon definition of effective cooperation in the literature on international waters (Huntjens et al. 2016; Zawahri 2008). Through a review of the literature and a qualitative content analysis of the literature and interview data, I present a general definition of effective cooperation: *Effective cooperation achieves equitable results or outcomes through a cooperative decision-making process that has trusting relationships with stakeholder participation and communication. This process is supported by the appropriate legal and institutional arrangements, where the outcomes can be the resolution of the initial problems, reaching the goals of the parties, or complying with mutually agreed-upon principles in a legal*

DOI: 10.4324/9781003126249-12

arrangement. While this definition does capture many of the aspects of effectiveness discussed in Chapter 8, it is not a measurable definition because "evaluating the 'effectiveness' of a cooperative arrangement means comparing something" (Underdal 1992). However, what we are comparing and what the results or outcomes should be is not defined. Defining effectiveness is "problematic in the sense that everyone is going to have their own definition of effective" as stated by Dr. Dave Devlaeminck of Chongqing University (Devlaeminck 2018).

This chapter aims to address these issues by answering the second part of the challenge addressed in Part III: how can we evaluate the degree of effectiveness in a transboundary cooperative process? This chapter develops the Weighted Model of Effective Transboundary Water Cooperation, an instrument for evaluating effective cooperation; in addition, it outlines a three-stage process for using the model.

Introducing the Weighted Model of Effective Cooperation

The Weighted Model of Effective Cooperation is based on the Four Frames of Cooperation described in Chapter 6. Cooperation is defined through the four frames and then evaluated through the model. The Weighted Model of Effective Cooperation can be visualized by adapting Underdal's stages of regime formation (see Figure 9.1).

Similar to Figure 8.1, Figure 9.1 shows the stages of development for a cooperative regime and an example of how the Four Frames of Cooperation can be applied. It differs from Figure 8.1 in that it shows more detailed interlinkages between the stages, as all three stages can exist simultaneously. Given the nature of transboundary waters, cooperation is not a linear process that results in an impact or outcome, but rather it is a process that must adapt, redefine the issues, develop new solutions, and strive for mutual benefits that are flexible with respect to change. Furthermore, this process occurs at a specific context and scale, which will determine what the appropriate elements and aspects of effective cooperation are. This model is based on the more traditional view of regime formation from international relations literature; however, what is missing from this conceptualization of effective cooperation is the emphasis on the relational and outcome frames that emerged in the data, as presented in Chapter 8.

The data analysis found that out of the four frames – legal, institutional, outcome, and relational – the most commonly referenced aspects of effective cooperation fall within the relational and outcome frames. Recall that the relational elements of cooperation are defined as the process and relationship building between all relevant stakeholders, including trust and shared understandings, while communicating transparently to identify and pursue mutually beneficial outcomes. The outcome frame is defined as the benefits, goals, actions, or specific outputs of cooperation that the relational frame is

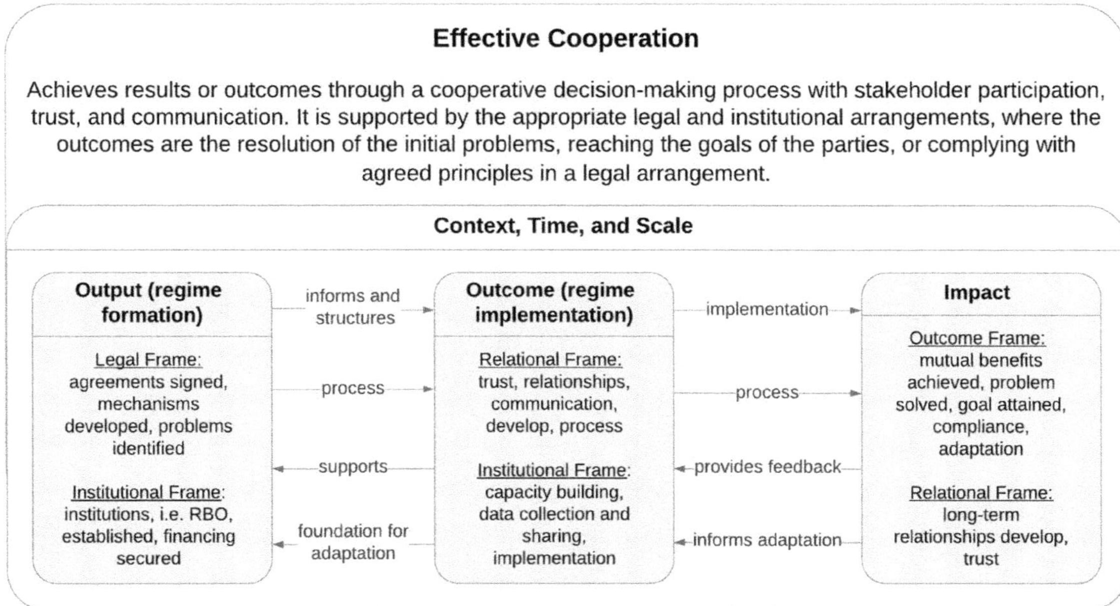

Figure 9.1 The Weighted Model of Effective Cooperation as shown through the stages of regime effectiveness adapted from Underdal 2002.

The figure shows the definition of effective cooperation and the frames of cooperation within the three stages. Cooperative processes are not linear, and the stages link together or even exist simultaneously, represented by the arrows connecting the stages.

The outcome stage is defined as the changes in human behavior as a result of the regime. This differs from the definition of the outcomes and outcome frame that are used within the Four Frames of Cooperation, which are the benefits, goals, actions, or specific outputs of cooperation.

attempting to identify and achieve. This finding supports the observation of several researchers that the evaluation of cooperation needs to extend beyond the signing of agreements and the establishment of institutions; these, while important elements of cooperation, do not guarantee effective cooperation. Daoudy and Kistin (2008, 2) presented this argument:

> Given such assumptions, the mere existence of cooperative arrangements is often celebrated as a sign of progress, with little or no interrogation of their influence on core issues related to transboundary water management (Kistin and Phillips 2007). Yet, as highlighted in academic research, the mere attempt to cooperate does not guarantee effects, let alone optimal outcomes (Helm and Sprinz 2000; Miles et al. 2002).
>
> (Daoudy and Kistin 2008, 2)

Therefore, I developed a different interpretation for modeling effective cooperation that embodies all Four Frames of Cooperation while putting a heavier weight on the relational and outcome frames. See Figure 9.2 for a representation of the weighted frames and how these are intertwined.

The outcome frame of cooperation is at the center of effective cooperation – represented by the largest darkest circle in Figure 9.2. Without achieving the benefits, goals, actions, or specific outputs of cooperation, how can we say cooperation is effective? Ms. Silvia Rafaelli of the Global Environmental Facility stated succinctly, "effective is when you see the results" (Rafaelli 2018). In order to achieve these outcomes, the cooperative process should incorporate appropriate aspects from the relational, legal, and institutional frames. What is deemed as 'appropriate' is dependent on the context and scale of the basin or aquifer. The relational frame is larger and darker in Figure 9.2 than the legal and institutional frames in this representation of the model. This incorporates the findings from Chapter 8, where the relational elements were the most commonly referenced aspects of effectiveness when interviewees were asked to define effective cooperation, as well as the common thread in the basins and aquifers they listed as examples of effective cooperation. In fact, one could argue that even if cooperation achieved its outcomes, if the trust, communication, and relationships were not there, it would not be considered a fully effective cooperative process. This differs from other conceptualizations of cooperation, where trust, confidence, and relationships are considered part of the enabling environment, rather than part of the cooperation itself (Grey, Sadoff, and Connors 2009). The other attribute that should be pointed out in the model is that the frames are not isolated – elements of cooperation and aspects of effectiveness overlap. Recall the example of mutual benefits overlapping between the outcome and relational frames discussed in Chapter 6. This is demonstrated by the lines in Figure 9.2 connecting all the frames together in an attempt to show how all four frames must work together, as appropriate for the particular basin or aquifer, for there to be effective cooperation.

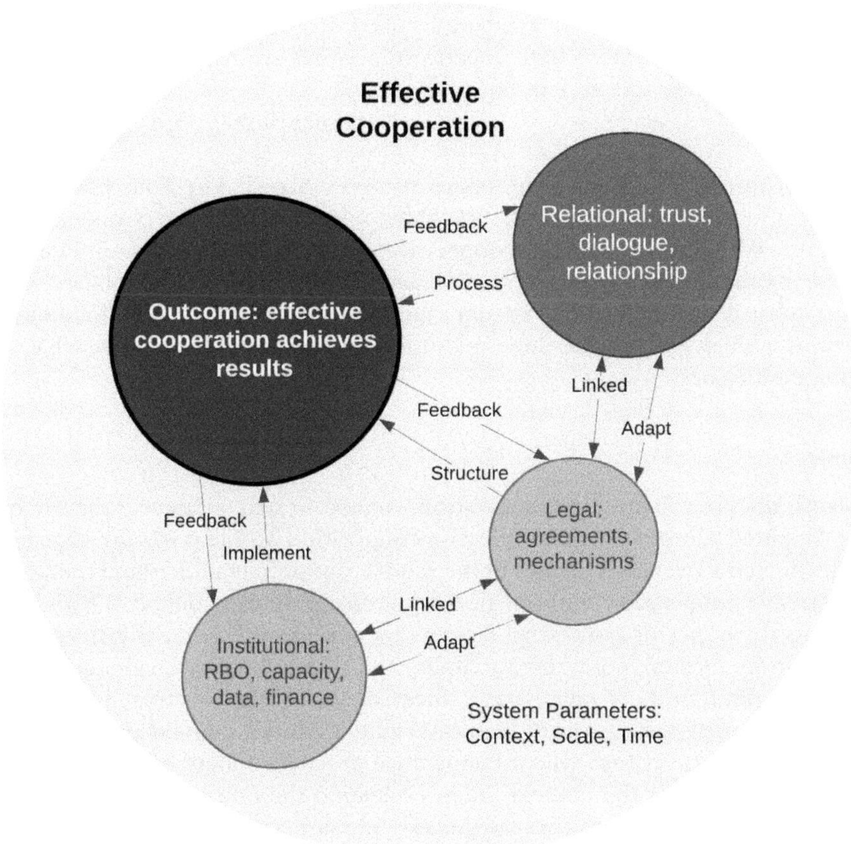

Figure 9.2 The Weighted Model of Effective Cooperation.

This figure shows the Four Frames of Cooperation and how the outcome and relational frames are weighted more heavily in the evaluation of effective cooperation. These all take place within the system parameters of context, scale, and time.

Expanding on the Weighted Model of Effective Cooperation: What is being evaluated?

These two graphics (Figure 9.1 and Figure 9.2) attempt to represent the conceptual framework for the Weighted Model of Effective Cooperation. However, to evaluate the effectiveness, there needs to be a comparison against a standard of success or accomplishment. The international relations literature on regime effectiveness has a consensus on what a model must answer in order to evaluate effectiveness. The model needs to answer three conceptual questions that are put forward in the literature (see Underdal 1992, 2002; Bernauer 1995):

> (i) What precisely constitutes the object to be evaluated? (ii) Against which standard is this object to be evaluated? (iii) How do we go about comparing the object to our standard; in other words, what kind of measurement operations do we have to perform in order to attribute a certain score of effectiveness to a certain object (regime)?
>
> (Underdal 2002, 5; 1992, 228–29)

The first question is: what is the object to be evaluated? The Four Frames of Cooperation is at the core of the Weighted Model of Effective Cooperation (hereafter Weighted Model). The cooperative 'system' defined in the Four Frames is the 'regime' or the object to be evaluated. However, several parameters will affect what is included in the 'system': context, scale, and time. The following sections address each of the three questions, beginning with the first: what is being evaluated?

Context

As with the Four Frames of Cooperation, context heavily influences the use of the Weighted Model to evaluate effective cooperation. Context sets the stage for evaluating effectiveness. The user of the model cannot fully understand the scale of effective cooperation without first acknowledging the context. Chapter 6 presents three specific contextual factors that appear in the data as influencing cooperation: meanings of water, variability and risk, and donor financing. These examples illustrate how context can affect the structure, function, and operation of a cooperative process. In the Weighted Model, context provides the constraints on the system where cooperative processes function; therefore, we need to be aware of the context when evaluating the effectiveness. Generally, we can group context into two categories – physical and human. Physical context includes geography, climate, hydrology, and so on – in essence, the ecological and environmental aspects of a basin or aquifer. Human context includes the politics, culture, history, economics, and so on – in essence, all the relevant social aspects in the basin or aquifer. By considering the physical and human contextual factors together, we can define the nature of the collective action problem that necessitated cooperation. The unique combination of these two factors will determine and influence the elements of cooperation that form; it also informs the evaluation as to whether they are the appropriate aspects or whether they hinder the development of effective cooperation. For example, two riparians are integrated regionally beyond water, and both states are willing and ready to enter into a cooperative process on transboundary waters. The willingness of the states in this example satisfies the relational frame and would be viewed more strongly than if the context of regional integration was not known. This is because we know from the literature that states are more willing to cooperate over transboundary waters if there is cooperation and regional integration over non-water issues (Boklan and Janusz-Pawletta 2017; Conca, Fengshi Wu, and Cigi Mei 2006).

In addition to the degree of complexity that context brings to the model, it also presents two additional challenges. Dr. Naho Mirumachi of King's College London noted that effective cooperation is "very context-dependent. What might look as effective cooperation might not translate in its exact form to another river basin. For me, searching for this universal outfit or this universal shape of effective cooperation might not be very productive or fruitful because of the local variability, because of the specific socioeconomic dynamics that are going on in these basins" (Mirumachi 2018).

This statement summarizes the key challenge with context and effectiveness. How can we define effectiveness if it is different for every basin or aquifer? Ultimately, this chapter develops a model for defining and evaluating effectiveness based on the context of the basin or aquifer. What is considered an element of cooperation or considered to be effective will differ from basin-to-basin and aquifer-to-aquifer. This means that whether or not a basin or aquifer is considered to have effective cooperation, it will not be comparable to another basin or aquifer. Instead, the model and its outcomes are meant to be used by actors to improve their cooperation to be more effective. The second challenge put forward by a different interviewee relates to how context impacts what is considered effective. Dr. Susanne Schmeier of IHE-Delft stated, "I just think that it depends so much more on the basin context than I thought back then, in both a positive and a negative way. On the one hand, of course effectiveness, whether someone in the basin sees basin organization or cooperation process as effective, will depend on what they expect from cooperation. But then what I also really came to acknowledge in the past years is that what people expect from basin cooperation is not necessarily what's best for the basin" (Schmeier 2018).

This statement has two key ideas: the first idea is that effectiveness is dependent on who is evaluating it. Therefore, we need to consider multiple viewpoints when evaluating effectiveness, as what is effective for the hegemon in the basin is likely not effective for the states or other actors with less power. The second idea is that defining effectiveness based solely on the perceptions of a single group, or perhaps from the contextual perspective of the basin, may not be what is needed for the basin. Effectiveness does not always equate to what is best for the basin. For example, if states agree to over allocate and fully consume the flow of a river and are able to do so, is that cooperation effective, as they have achieved their goals? Therefore, we need to consider context-based 'objective' standards for evaluating effectiveness, in addition to standards defined by the parties involved. This is discussed in more detail in the following sections.

Context has the ability to shape and transform the structures and functions of cooperation. It impacts what key elements are represented in the Four Frames of Cooperation, how these are perceived to be effective, and what standards we are comparing them to in order to evaluate the extent of effectiveness. While context may seem obvious, it is a critical component of the Weighted Model of Effective Cooperation.

Scale

As with the Four Frames of Cooperation, the Weighted Model of Effective Cooperation is impacted by scale. Scale is linked to context, as context will shift depending on the scale under consideration. It could even be considered an aspect of the greater context. However, scale deserves special consideration with regard to context. In Chapter 6, scale is defined as the territorial delineation of a biological or ecological landscape or of a social, political, or economic entity, which incorporates the spatial dimension into transboundary water cooperation. This definition remains consistent in the Weighted Model. Scale can be categorized into two general categories – the geographic and the governance scales. The geographic scale is the geographic extent of nested basins and aquifer areas corresponding to the geographic scope of transboundary water cooperation. The governance scale embodies political geography's understanding and inclusion of power: the international system, supranational organizations, nation-states, regional governments, city, community, and local scales. See section on Scale in Chapter 6 for a full discussion of the two categories of scale. The two scales have been subdivided into various levels in order to create a common language for discussing scale; however, the model recognizes that scale is context-dependent and can operate at different levels. Figure 9.3 illustrates the subdivisions in the governance and geographic scales.

As mentioned in Chapter 6, these two scales are often perceived as linked, where the larger the geographic scale, the higher the governance scale is required, as more power is needed to manage or cooperate. This state-centric view equates larger size with higher power in the hierarchical view of scale

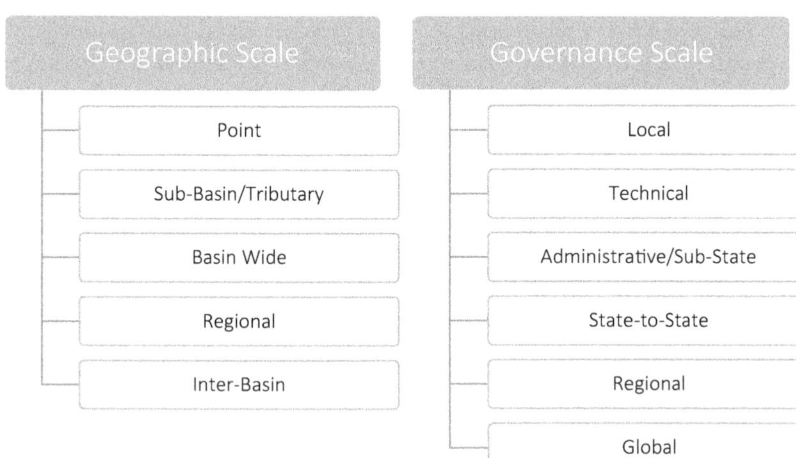

Figure 9.3 Geographic and governance scales.

These create a common language for discussing scale; however, they should not restrict other conceptualizations of scale.

(Graddy 2011). As with the Four Frames of Cooperation, the geographic and governance scales should not be considered as linked in the Weighted Model; instead, they should be evaluated separately with respect to their impact on the effectiveness of cooperation. How the governance and geographic scales are defined for a particular shared water will affect the format, function, and ultimately, the effectiveness of a cooperative process.

As a part of the qualitative content analysis, the expert interviewees were asked: At what level does cooperation need to occur to be considered effective? In the 47 responses to this question, there are several common themes, as well as varied and nuanced answers. In general, most of the interviewees stated that cooperation at all governance scales was necessary – the governance scale was also the more commonly mentioned out of the two scales. Others highlighted that specific scales were particularly important. A few mentioned that the smaller governance scales, such as the local and community levels, were important as these are the scales where many of the objectives are implemented and impacts are felt. More interviewees, however, mentioned that the national or state-to-state governance scale was the most important for effective cooperation. The national governments need to be involved in creating general principles, politically backing cooperation and allowing it to occur at smaller scales, and establishing a larger governance mechanism that considers the entire basin. Even when mentioning that a particular scale was the most important, everyone recognized that governance at all scales is needed. This was often stated with a caveat, that it is dependent on the context and what issue the cooperation is focused on. It is not only context that impacts the question of scale and effective cooperation. Dr. Naho Mirumachi of King's College London pointed out that much of the focus historically has been on the interstate scale; "looking at the state level is not sufficient. … There are a lot of synergies that are at play from the domestic, from the subnational to the international. … I think the keyword here is polycentric, that there is no one scale that determines the process of cooperation or the outcomes of cooperation, that there are different spatial scales that drive the process" (Mirumachi 2018).

Similarly, one interviewee expressed that scales are linked and should not be put into silos; therefore, for cooperation to be effective, dialogue and cooperation are needed at various levels that extend both vertically and horizontally. The Rhine and Danube (along with the Sava River sub-basin) were given as examples of multiscalar governance, where there is a basin-wide commission, with smaller sub-basin commissions and the support of national organizations and policy. Despite asking about the geographic scale in particular, in many interviews, the conversation shifted back toward the governance scale or discussed the two in concert with each other. For example, Dr. Dinara Ziganshina of the Scientific Information Center of Interstate Commission for Water Coordination in Central Asia noted that at the basin-wide scale, general principles should be established by the national governments, which would then trickle down to enable cooperation at lower scales, as these are the scales where people are dealing with the main issues that necessitate cooperation on a day-to-day basis (Ziganshina 2018).

In addition to context, the scale of governance and the geographic scale will vary depending on the problem or issue at hand. This can also lead to a misalignment between the appropriate scale and the issue to be addressed. Problems over shared water resources can manifest at different geographic scales and require different governance scales to address issues. Several interviews noted that when identifying the appropriate governance scale to resolve a problem, it should match the issue at hand. For example, Dr. Zach Sugg of the Gould Center for Conflict Resolution stated, "My whole thinking about that is that the governance and cooperation arrangements should be matched appropriately to just the nature of those problems. Starting from the principle of decentralization, building up political scales that are matched, in an ideal world, matched to the resource and spatial scale, too" (Sugg 2018). The experts with law backgrounds, in particular, put forward that determining the appropriate scale to address an issue should be driven by subsidiarity. The principle of subsidiarity is the idea that decision-making authority should be located at the level of accountability where the outcomes will occur and near the location where the actions to achieve the outcomes will happen (P.J. Wolf 2001). Stated simply, the cooperative process should take place at the lowest appropriate level – e.g., global issues addressed at the international scale, regional problems at the regional scale, basin-wide issues at the basin-wide scale, and so forth (Benson and Jordan 2015). Determining what the lowest appropriate level is has been challenging and may be dependent on "everything from geography to capacity to local knowledge," according to Dr. Gabriel Eckstein of Texas A&M School of Law. However, despite the challenge, determining what the appropriate level is has been successful in many places, particularly within Europe (Eckstein 2018).

With respect to the Weighted Model, there is no appropriate scale that can be predetermined as the most 'effective' without considering the specific context of the basin and the issue at hand. Therefore, when using the model, the scale of meaning needs to be defined. The scale of meaning is the geographic and/or governance level at which a problem is experienced, analyzed, discussed, or relevant, given the context of the problem and the place (Lebel, Garden, and Imamura 2005). If the scale of meaning is established, this can be compared to the scales at which the problem is being addressed, which may offer a perspective on the extent of effectiveness in the basin or aquifer. The scale of meaning could then aid in applying the principle of subsidiarity to cooperative processes and assist in matching the scale to the problem. Ultimately, the user of the Weighted Model of Effective Cooperation needs to clearly define the problem or process of interest, its geographic scale, and the governance scale addressing the issue to fully define the system of cooperation that is being evaluated through the model.

Time

Like scale, time could be considered a component of context; however, I believe it is essential to separate time. Time will impact the context of a basin or aquifer and what key elements are present and the extent to which a process has had time to function or implement policy into outcomes. The question of time and effectiveness came up in 17 of the interviews. Ultimately, three overarching considerations related to time emerged. The first is the most obvious. It takes time for cooperation to evolve, treaties to be negotiated, trust to develop, policy to be implemented, relationships to mature, capacity to be built, and more. Actors must be willing to invest time into the cooperative process, as time is required for cooperation to evolve into effective cooperation. Second, when in time is the object (the cooperative process) being defined? Are we considering a point in time with a specific scale and context, or are we considering a set time period, or effectiveness overall? These will each yield a different result regarding the effectiveness of cooperation in a basin or aquifer. Generally, the Weighted Model is constructed to evaluate effectiveness at a certain point in time while still recognizing the history and the future goals defined for a basin or aquifer. However, if the user wanted to consider the effectiveness over a time period or overall, they could evaluate the effectiveness at multiple points in time, using the same scales, and compare them. This might help point to attributes or contextual factors that increased or decreased the effectiveness of the cooperative process. It is important to remember that cooperation ebbs and flows, and while it might appear ineffective at one point in time, it does not mean that cooperation overall has been ineffective – context is essential. This same question can be asked in terms of where in the development process is cooperation at this point in time? Are we evaluating cooperation during the negotiation of an agreement, right after an agreement is signed, or ten years after implementation? I would argue that all of these stages are part of the process of cooperation and the level of effectiveness at that point in time and should, to a degree, reflect where the cooperative process is. For example, if we are considering a basin that is negotiating an agreement at the point in time of evaluation, we would not expect the negotiated goals of the agreement to be achieved. However, we could evaluate the level of relational elements, as well as aspects of the legal and institutional frames. In this respect, the Weighted Model is modular, as it can be adapted based on the definition of cooperation for the place at a specific point in time through the use of the frames of cooperation.

The last consideration is whether outcomes or particular elements have been achieved within a set period of time or a timeline defined by the cooperative process. This is a tricky component, as how long it takes to achieve an outcome is heavily dependent on the outcome as well as the context. Depending on the context of the basin or aquifer and the particular outcome itself, this could be used as a measure of whether a process is effective; however, it may reflect more on the efficiency of the process rather than the effectiveness. Considerable care should be taken when using the time to achieve a goal as a way to determine effectiveness, particularly without having an understanding of the context involved.

Similar to context, time also brings challenges to comparability. Dr. Susanne Schmeier at IHE-Delft highlighted that the rate of development of a cooperative process in time is different for every basin or aquifer and even in the same basin or aquifer at different points in time, depending on the internal and external factors contributing to cooperation (Schmeier 2018). Therefore, the Weighted Model is not developed to compare progress over time between places because context impacts the rate and type of development of cooperation. For example, cooperation in the Lake Chad basin was established in 1964 but was not functioning for several decades until the Lake Chad Commission was revived in the 2000s. Whereas, the Orange-Senqu River Basin established cooperation in the 2000s and has had a significant amount of progress. Superficially, one might be inclined to say that cooperation in the Lake Chad basin was ineffective in that it ceased to function, while in the Orange-Senqu basin it is effective in that it has progressed rapidly over a short period of time. However, without taking the context – such as the internal and external and human and physical factors – into account, this is essentially comparing apples to oranges. Ultimately, time – like context and scale – is a complexifier. The user of the Weighted Model needs to take time into consideration when evaluating effectiveness and be clear as to when in time the cooperative process is being evaluated.

In addition to setting the parameters – context, scale, and time – of the cooperative process that is being evaluated, the user must also be aware of which stage of the cooperative process is being evaluated and what data is being used. For example, the international relations literature has identified *output*, *outcome*,[2] and *impacts* as stages in the evolution of a regime, as shown in Figure 9.1 above. *Output* is the regime formation – the signing of an agreement; *outcome* is the change in human behavior, and *impact* is the responses to the changed behavior (which is captured by the Outcome Frame in the Four Frames of Cooperation). When discussing effective cooperation, as can be seen from the data in Chapter 8, cooperation is considered effective when there is an impact. However, much of the literature that aims to measure cooperation evaluates the output, that is, whether an agreement has been signed, the existence of a river basin organization, or specific mechanisms in a treaty. It is difficult to measure impact until after the fact – after the results of the cooperative process have been achieved. Therefore, there needs to be caution in measuring impact effectiveness while using output data (Underdal 1992). In terms of the Weighed Model of Effective Cooperation, the effectiveness of each stage will be addressed separately through the examination of each of the Four Frames. The object being evaluated, that is, elements within each frame, will be compared with the data within each frame. Now that we have established how the object being evaluated is defined by its parameters – the context, scale, and time – the standard to which we are comparing the cooperative process must be determined.

Expanding on the Weighted Model: What is the evaluation standard?

As mentioned above, to determine whether a cooperative process is effective, it must be compared to some standard that defines success or failure. This is the idea of the second question posed by Underdal (2002): "Against which standard is this object to be evaluated?" Establishing this standard is challenging, as it is subjective and variable depending on the goals of the cooperative process and the goals of the person or actor evaluating the cooperation. For example, many of the indicators developed to measure transboundary water cooperation focus on procedural aspects, such as the existence of a treaty, regular meetings, allocation mechanisms, or dispute resolution mechanisms in an agreement or institution. These evaluations are often derived from international water law, such as the provisions put forward in the conventions and customary law, for example, the criteria used to measure arrangements with operational cooperation in SDG Indicator 6.5.2. (see Chapter 3). In the regime effectiveness literature (see Chapter 7), two main steps are put forward for determining the standard for comparison: (1) what is the point of reference that the object will be compared to and (2) what is the unit of measurement (Underdal 1992, 2002).

Point of reference

Points of reference are used to determine improvement created by the regime. The point of reference is defined in two ways. The regime can be compared to the collective optimum – or the ideal – which is defined as the solution that "accomplishes, for the group of members, all that can be accomplished -given the state of knowledge at the time" (Underdal 2002, 8). The second option for the point of reference is the counterfactual, which can either be a no-regime counterfactual or a baseline counterfactual (Young 2011; Underdal 2002, 1992). The Weighted Model of Effective Cooperation provides space for comparison between both the counterfactual and the ideal depending on the problem or issue of interest and the subjectivity of the evaluator.

The first option in the Weighted Model, the process of interest, as defined by the parameters, is compared to the ideal scenario – or the collective optimum. Questions will be asked to the actors to develop the ideal scenario given the basin or aquifer context; this will establish the key elements and aspects of effective cooperation in the ideal scenario. These can then be compared to the key elements and aspects of effective cooperation in the process of interest at the point in time that is being evaluated. It will also allow for the comparison of the ideal scenario between actors, which could inform future negotiation or adaption of the cooperative process.

The second option in the Weighted Model is to compare the process of interest to the counterfactual. In the literature, the no-regime counterfactual is defined as the state of affairs if the cooperative regime had never existed (Young 2011; Underdal 2002). When compared to the no-regime condition, the outcome is a measure of effectiveness in absolute terms. I do not believe that the comparison to a no-regime condition is helpful to most basins or aquifers. In the case of most surface water, there are generally some history, past agreements, cooperation, or conflict between riparians in other sectors that will influence the development of cooperation over water. Therefore, a no-regime counterfactual is an unrealistic scenario, and the measurement of progress toward effective cooperation would be less applicable to basins and aquifers aiming to measure their actual progress in developing effective cooperation. The caveat is with groundwater; in some aquifers, there may be a no-regime situation occurring, as there may be no history of states cooperating over groundwater. The Weighted Model of Effective Cooperation is designed for comparison with the baseline counterfactual. In the literature, the baseline counterfactual is the hypothetical situation if the previous rules, norms, and principles had remained in place (Underdal 2002, 1992). This aims to measure the relative or incremental improvement in the effectiveness of the cooperative process, which should be more relevant and applicable to actors implementing this model. However, rather than comparing the model to a hypothetical scenario where the baseline is extrapolated to the current point in time, the Weighted Model proposes that the baseline condition is an actual previous point in time that can be compared to the current point in time as defined in the parameters. Using this method, we can compare the point in time of interest to multiple previous baselines and track relative improvement and the rate of progress over time in the basin or aquifer for the issue under consideration. Therefore, in the model, the actors are asked questions with the purpose of eliciting answers that define, describe, and identify key elements and aspects of effective cooperation at the baseline and at the point in time of interest being evaluated. These can then be compared to determine the relative progress in developing effective cooperation.

Unit of measurement

In addition to the point of reference, a standard metric is needed for evaluating effectiveness. Determining the unit of measurement is one component of effectiveness evaluation where the subjectivity of the evaluator can have a significant impact. For example, Dr. Francesco Sindico of Strathclyde Law School, when asked to define effective cooperation, stated, "What kind of effectiveness are we talking about? Is it environmental effectiveness? Is it political effectiveness? Is it legal effectiveness?" (Sindico 2018). Each of these types of effectiveness will have a different perspective on what key elements or aspects of effectiveness are involved in transboundary water cooperation. Effectiveness depends not only on the context of the basin but the individual perspective or goal in measuring effectiveness. For example, in the early to mid-1900s, when the United States was at the peak of dam and large infrastructure construction, cooperation would have been considered effective if all the water was allocated and used for economic development and growth. These outcomes would not be considered effective by today's standards in the United States because they purposefully ignore ecosystem health and environmental flows. In the regime effectiveness literature, Underdal (1992) proposes that a standard metric for the measurement of regime effectiveness could be human welfare or ecological properties. Young and Levy (1999) define five different approaches to effectiveness that use different standards: problem solving, legal, economic, normative, and political. The idea of defining a standard metric for effective cooperation relates to the discussion in Chapter 6 on the neutrality of transboundary water cooperation as defined in the Four Frames of Cooperation. In the Four Frames of Cooperation, transboundary water cooperation is strictly defined as having no value judgment on cooperation. By evaluating the effectiveness of cooperation, a value judgment is made by the Weighted Model based on the standard metric of comparison. This model is constructed to value effective cooperation as cooperation that is equitable and sustainable, based on current thinking in international water literature and global sustainable development efforts. Dr. Mark Zeitoun, at the University of East Anglia, explains the neutrality of cooperation and the need for a goal when making a value judgment stating, "When I say cooperation now, I'm not making any value judgment. I don't say cooperation is good or desired, but certain types of cooperation are definitely necessary for the goal – the goal being a sustainable and equitable transboundary water arrangement. The goal is definitely not cooperation. Cooperation is just a means to an end, and [it] can be constructive or destructive" (Zeitoun 2018). Therefore, with the Weighted Model of Effective Cooperation, I am advocating that effective cooperation is equitable and sustainable. Equitability and sustainability are the two units of measurement for comparison. These two concepts have a history in the global policy and research sphere, with the work of the London Water Research Group bringing them together in the context of transboundary water cooperation. For example, Zeitoun et al. (2011) write, "A second important step is to refine, focus and substantiate the meaning of "transboundary water cooperation." Both this paper (Zeitoun, Mirumachi, and Warner 2011) and the one it builds upon (Zeitoun and Mirumachi 2008) have highlighted the less benign faces of cooperation and suggested that the term be applied more selectively to reflect equitability and sustainability."

While it may seem intuitive or logical to select equitability and sustainability as the units of measurement for effective cooperation, I want to briefly summarize why these aspects were selected over other potential units, such as water security. Both of these terms are prominent in the literature and the discourse in the field of international waters and in the larger institutional system. Equitable is most commonly referenced in discussions of the principle of equitable and reasonable use. Articles 5 and 6 of the *1997 UN Watercourses Convention* describe how states should share water in an international watercourse with "a view to attaining optimal and sustainable utilization thereof and benefits therefrom, taking into account the interests of the watercourse States concerned, consistent with adequate protection of the watercourse." Article 6 outlines factors for consideration when determining the uses of water that are reasonable and equitable (United Nations 1997). Therefore, equitable is most often discussed with respect to allocation and utilization, where "the principle of equitable utilization emphasizes that a state cannot legally do as it pleases with the transboundary water resources in its territory. Its essence is that states must act reasonably in dealing with these waters" (Bourne 1998, 216). However, in the

Weighted Model, equitable extends beyond allocation and use to other facets of transboundary water cooperation. Research by the London Water Research Group has shown that not all cooperation is inherently good, nor is all conflict bad. Cooperation can maintain the status quo that enshrines the power of the hegemon in the basin in inequitable agreements and outcomes (Zeitoun and Mirumachi 2008; Zeitoun, Mirumachi, and Warner 2011; Zeitoun et al. 2017). This supports the need for consideration of equitable measures of effectiveness in all Four Frames of Cooperation. It is worth clarifying that in the Weighted Model, equitable and equal are not the same. Equity or equitable refers to fairness or justice in the cooperative process of sharing transboundary waters. Equitability can be perceived to be achieved when there is an absence of disparities between the actors in the basin in both outcome and process (and perhaps between sectors of use). For each unique basin or aquifer context, equity will be perceived differently.

Like equitability, sustainability has become prominent in international discourse. Sustainable development, in particular, has gained increased focus since the 1980s with the publication of the UN report in 1987, *Our Common Future*. This report defines sustainable development as "development that meets the needs of the present without compromising the ability of future generations to meet their own needs" (Brundtland Commission 1987, 41). Sustainability and sustainable development are often used interchangeably in the literature, even though there are calls for decoupling the terms, as sustainable development is economic growth-centered, which may further entrench western growth and capitalist ideals (Purvis, Mao, and Robinson 2019). The most common view of sustainability or sustainable development is the three pillars conceptualization, see Figure 9.4, in which sustainability is held up "as the balancing of trade-offs between seemingly equally desirable goals with these three categories" – social, economic, and environmental (Purvis, Mao, and Robinson 2019, 5). Others define sustainability as "a condition in which natural and social systems survive and thrive together infinitely (Euston and Gibson 1995)" (Svendsen 2005, 5). Given the prominence of sustainability in the SDG Framework and the fairly global acceptance of sustainability through the foundational social, economic, and environmental pillars, the Weighted Model uses this framework as the other standardized unit of measure. While the appropriate social, economic, and environmental elements of cooperation will vary between basins, these pillars are used as a means to create 'objective' measures of effective cooperation that look beyond the outcomes defined by the parties involved.

As shown through this discussion as well as in the literature review on effectiveness in Chapter 7, establishing a standard metric for comparison is challenging and subjective. Based on global discourse on sustainability and equity in the international waters literature, the Weighted Model of Effective Cooperation uses equitability and sustainability as the standard unit of measure. This culminates in the statement that *effective transboundary water cooperation is equitable and sustainable*.

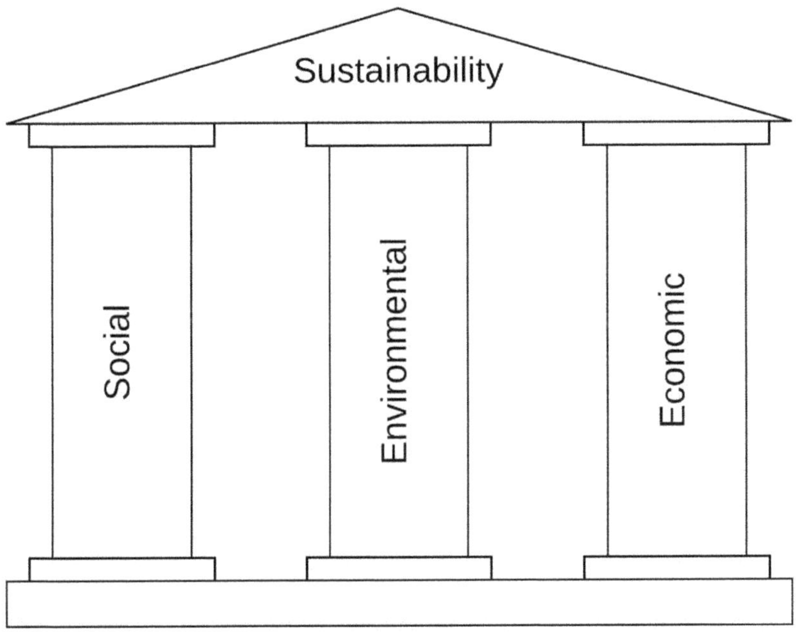

Figure 9.4 The three pillars of sustainability and sustainable development.

Expanding on the Weighted Model of Effective Cooperation: How do we compare against the standard?

The final question is, how do we compare the cooperative process to the standard, either the baseline or the ideal, using equitability and sustainability as the units of measurement? This is a challenging question that is currently being worked through in the literature (see Miles et al. 2002 for a discussion and example of operational measurement of regime effectiveness). Establishing an empirical method to analyze the data on effective regimes is difficult, given that most methods are comparing the current regime to either the hypothetical ideal or a hypothetical baseline where empirical data is not available. In addition, it is challenging to establish causal links between the regime, changes in behavior, and the resulting impacts (Underdal 2002). Therefore, the Weighted Model presents a method for evaluating effective cooperation but does not claim to establish or determine any causal links between the key elements or aspects of effectiveness with effective cooperation.

The most common methods to measure effectiveness are goal attainment, problem solving, and compliance. Daoudy and Kistin (2008) argue for an additional method that is based on a power analysis of the cooperative process. This interest-based method aims to highlight issues with cooperative processes maintaining power imbalances, which helps to address the standard for equitability.

Goal attainment focuses on "the extent to which a regime's (stated or unstated) goals are attained over time" (Young 1994, 144; Schmeier 2013). Ultimately, it asks how the actors are meeting the joint goals, expectations, and objectives they set. There are three main critiques of this method. First, the goals defined by the parties may not be appropriate for the basin (Daoudy and Kistin 2008). In the Weighted Model, an attempt is made to correct this by including standardized 'objective' questions that reflect the three pillars of sustainability and whether these are addressed in the cooperative process. The second critique is that focusing on goal attainment may distract from distributional, relational, and unintended effects (Young 2001). Lastly, this method is challenging to measure, as goals are often vaguely defined or immeasurable (Schmeier 2013; Daoudy and Kistin 2008).

Problem solving is when the issue or problem the regime was established to solve has been eliminated or alleviated (Young and Levy 1999). This method is often seen as the true measure of effectiveness in terms of whether a regime has failed or succeeded. However, measurement generally results in failure to resolve the problem, particularly in transboundary waters, where the problems are not usually solvable and instead need to be managed. Focusing on problem solving neglects the effort or commitment that parties put into solving the problems at hand (Daoudy and Kistin 2008). In addition, the problems necessitating cooperation on international waters are often not quantifiable.

Compliance is the extent to which the cooperative process and the actors conform to the established rules, such as adhering to water allocations (Daoudy and Kistin 2008). The critique of this method is similar to that of goal attainment in that the rules actors are complying with may not be appropriate for the basin. In some basins, there could be high compliance, such as for exchanging data and information, which may require very little change in behavior by the actors. Even though there is high compliance, there is potential for limited impact, no impact, and even negative impact on the shared water (Miles et al. 2002).

The final method proposed by Daoudy and Kistin (2008) is interest-based, which considers goal attainment for an actor's individual goals, as compared to the joint goals for the basin or aquifer. The power imbalances between actors can substantially impact the joint goals developed; therefore, this method considers whether agreements address the goals of all the actors' interests or just the goals of the more powerful actors (Daoudy and Kistin 2008). In the Weighted Model, this method is used to help establish the degree of equitability across the Four Frames of Cooperation.

In the Weighted Model of Effective Cooperation, these four methods for comparing the cooperative process to the standards are used in order to address the gaps and critiques in any one methodology. In addition, the best method to evaluate may vary depending on the context and the nature of the cooperative process itself. See Daoudy and Kistin 2008, Bernauer 1995, and Dombrowsky 2007 for examples where the evaluations have combined two or more of these methods to assess multiple aspects of the regime. The results of the evaluations of the four methods can then be aggregated into an ordinal scale ranging from lacking effectiveness to effective cooperation. Attributing a score in a higher order of measurement is beyond the scope and capacity of the Weighted Model, given the complexity of evaluating effective cooperation. Furthermore, caution should be applied when attempting to attribute a more precise score for effective cooperation using this model, as it may encourage comparison between basins and aquifers. Since cooperation and effectiveness are heavily dependent on context, scale, and time, and the assessment of the standard metrics is variable with the context of the basin, the extent and level of effective cooperation is not comparable between basins or aquifers. Wolf (2010, 108) sums this up by stating that "achievable international cooperation is always a long and complex journey, for which there is no single path and few shortcuts. Instead, there are many routes that can be followed and many steps that can be taken, with various options to consider and choices to be made."

How to use the Weighted Model of Effective Cooperation?

The Weighted Model of Effective Cooperation is a preliminary method for evaluating effective cooperation and measuring progress. It is a detailed method that incorporates the nuance and contextual richness of each basin and aquifer, meaning a comparison between basins and aquifers is not recommended. It is built on the theory and findings of regime effectiveness in international relations while incorporating expert knowledge from academics and practitioners in the transboundary water field. The above discussion in this chapter presents a conceptual graphical representation of how the frames of cooperation are weighted and interlinked while working within the system parameters. It further addresses the considerations that were taken into account when constructing the model. However, the goal of this book is to create a useful method to evaluate cooperation that can be applied, rather than strictly discussed in a theoretical or abstract sense. This section presents a step-by-step method for how to evaluate effective cooperation in a basin or aquifer using the Weighted Model. A multi-page graphic flowchart illustrates and summarizes this process, see Figure 10.3 in Chapter 10. The process for evaluating effective cooperation using the model is broken into three stages: parameters, assessment, and evaluation.

Stage 1: Parameters

Before reading through the first stage of the model, recall the two general definitions developed in this book.

1. *Transboundary water cooperation*: interactions between actors over shared waters that result in establishing mutually beneficial outcomes through a decision-making process; this process could include formal and informal,

legal and institutional mechanisms depending on the scale and context. (See Part II.)

2. *Effective transboundary water cooperation*: achieves equitable results or outcomes through a cooperative decision-making process that has trusting relationships with stakeholder participation and communication. It is supported by the appropriate legal and institutional arrangements, where the outcomes can be the resolution of the initial problems, reaching the goals of the parties, or complying with agreed-upon principles in a legal arrangement. Effective cooperation is a process that is equitable and sustainable. (See Part III.)

These definitions are general and provide concise definitions for broadly understanding what the model is considering. Keeping these in mind can help the user to determine the parameters of the cooperative system. As discussed, a cooperative process must be established in order to be evaluated. I believe a systems perspective is useful in conceptualizing the parameters that define what is or is not involved in the cooperative process that is under evaluation. The cooperative system is what is under evaluation, and the parameters establish the limits of the system. The parameters include the process of interest, scale, context, time, and actors. Figure 9.5 on the following page presents a pathway for determining each of these parameters. The first parameter to define is the process of interest. This is the problem that the cooperative process is aiming to solve or the particular aspects of cooperation between actors or within a river basin or aquifer that requires evaluation. For example, the process of interest could be the cooperation between states on the allocation of a particular river, the drilling and pumping from test wells in the border region of an aquifer, or even the basin-wide multi-sectoral management of a shared river. Without understanding the cooperative process or aspect of a process that must be evaluated, it is challenging to define the other parameters further.

Second, the user must identify the scale that this process of interest is occurring at. This parameter requires defining the current scales – geographic and governance – at which the process is occurring or defining the limits to the scales needed for evaluation. For example, if the process of interest is basin-wide multi-sector management, then the user may only be interested in the global, regional, and state-to-state governance scales and the basin-wide geographic scale. The user must also define the scale of meaning for the problem, as this has the potential to differ from the scales at which the process is occurring. When determining the scale of meaning, the principle of subsidiarity should be used to help establish the appropriate scale that the problem should be addressed at. Recall that the principle of subsidiarity calls for the decision making or cooperative process to be at the lowest appropriate scale. In this model, I have separated geographic scales into point, sub-basin, basin-wide, regional, and inter-basin categories, while the governance scales are

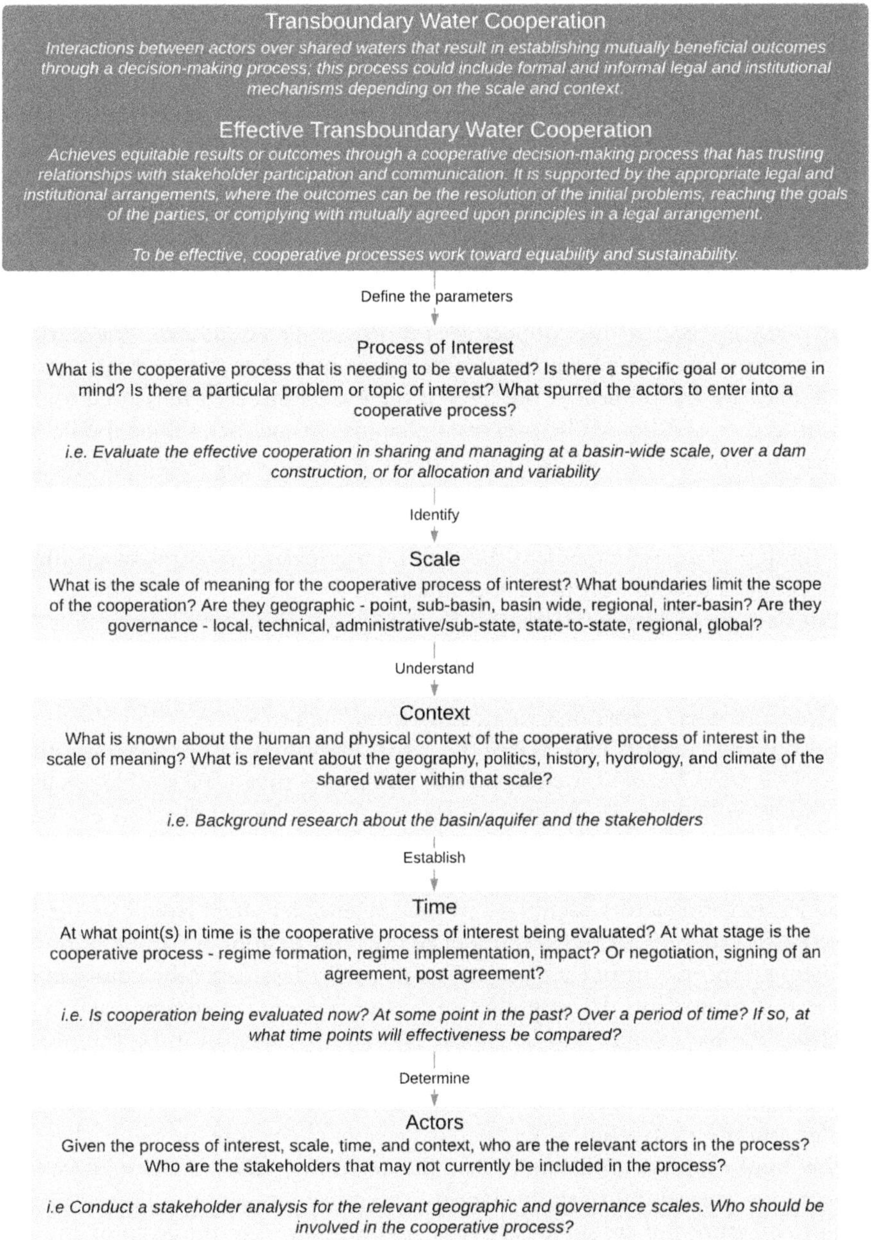

Figure 9.5 Stage 1: Parameters of the Weighted Model of Effective Cooperation.

local, technical, administrative/sub-state, state-to-state, regional, and global, see Figure 9.3. These are suggested categories to simplify discussions over scale, and additional or altered categories can be established provided they are relevant to the basin/aquifer and the process.

The third parameter that must be defined is context. It is necessary to better understand the context of the basin or aquifer, particularly with respect to the process. Background research should be conducted to identify the human and physical contextual elements that are relevant and influential to the process of interest and at the geographic and governance scales identified. This information will help inform and support the assessment and evaluation of the cooperative process. Following context, the timeframe must be established. At what point in time is the cooperative process being evaluated – the current point in time or some past point in time? This may alter the scale or context that is relevant. In addition, if the goal of the evaluation is to measure progress over time, the user should also identify the point in time that will be used as the baseline for comparison. Related to context, it may also be useful for the user to identify the stage the cooperative process is in at the point in time of the evaluation, such as regime formation, implementation, impact, negotiation, or post-signing of an agreement, as this will be particularly relevant when measuring progress toward effective cooperation.

The last parameter to define is the actors relevant to the process at the identified scale and time. The information gathered about the background and context of the process of interest should be used to identify what actors are involved and what actors may be missing from the process. A stakeholder analysis should be conducted to verify the actors identified. This is a critical piece of the Weighted Model for two reasons. The first is that a stakeholder analysis can ensure equitability in the evaluations that use the model. If actors excluded from the process are excluded from the evaluations that use the model, the inequity in the basin or aquifer is likely to be missed or understated. The second is that the actors identified in the stakeholder analysis will be used as primary data sources for the assessment of the Four Frames of Cooperation. By collecting responses from each actor, we can better determine the equity in the process by comparing the interests, goals, and outcomes between each actor. Defining the parameters may seem like an extensive step in using the model; however, it is vital to developing a measure of effectiveness that genuinely incorporates context into both the definition of cooperation for the basin or aquifer and the evaluation of its effectiveness. Furthermore, previous chapters in this book have demonstrated that the relational and outcome frames of cooperation are central to understanding effective cooperation. These are difficult to measure and are subjective relative to the actors in the basin or aquifer. Without this in-depth understanding prior to evaluating cooperation, the user might be limited to considering strictly legal or institutional design elements, which may not be implemented, complied with, or may not result in the intended outcomes.

Stage 2: Assessment

Once the process of interest has been identified, the scale(s) identified, the context explored, the points in time established, and the stakeholder analysis completed, the assessment of the Four Frames of Cooperation can be conducted. These parameters not only define the limits of the system or process to be evaluated but will also inform the detailed construction of the assessment instrument. This is the first step in the Assessment stage; Figure 9.6 presents a visualization and synopsis of this stage. To begin developing the assessment instrument, the evaluator must determine whether the goal is to measure progress from the baseline or to compare progress to the hypothetical ideal. This will inform the nature of the point of reference for comparison. Using this and the information gathered in Stage 1: Parameters, questions can be developed that are specific to the process of interest, context, and scales related to the baseline or the ideal. Related questions about the point in time under evaluation (labeled Current Point in Time in Figure 9.6) must also be developed. The responses from these two sets of questions are the data that will be compared in Stage 3. Questions for each frame of cooperation for both the baseline/ideal and the current point in time must be developed. These questions should be based on the context and the goals, outcomes, procedures, and so on that are established by the actors.

However, effective cooperation is not just measured by the elements defined by the actors. As seen in the literature and interview data, we know that the key elements or outcomes established by the actors in a cooperative process may not necessarily be the best for the basin or aquifer. Therefore, in addition to the basin/aquifer-specific questions, there should be standardized questions to bring relative objectivity to the assessment of the cooperative process that focuses on the units of measurement, which are equitability and sustainability. For example, research has shown that having a dispute resolution mechanism in a legal agreement increases the resilience of the agreement and provides a mechanism to resolve conflicts, thereby reducing the likelihood of conflict between riparian states. Therefore, a question such as 'Is there a dispute resolution mechanism?' could be a potential standardized question in the legal frame. This question development is particularly important for the outcome frame. As can be seen in Figure 9.6, there are two subsections of questions for both the baseline/ideal and the current point in time; these are the party defined and the standardized questions. The party-defined questions are the results, benefits, and outcomes as defined by the actors. In the outcome frame, the standardized questions capture the intent of equitability and sustainability. Questions should be refined to the specific context but should also ask about the achievement or existence of relevant goals within the three pillars of sustainability – social, environmental, and economic. For example, let us consider the cooperation between the United States and Canada on the Columbia River. The 1964 treaty's primary goals were to achieve hydropower energy production and flood control benefits.

Figure 9.6 Stage 2: Assessment of the Weighted Model of Effective Cooperation.

These are the party-defined outcomes, as stated in the treaty. The standardized questions, in this case, might ask, under the environmental pillar of sustainability, whether outcomes related to environmental flow, ecosystem health, or fishery protection are included or achieved in the basin. In this brief example, we can see that social and economic aspects have been included in the agreement with the party-defined goals of hydropower and flood control. Figure 9.6 provides a sample of general non-context-specific questions that can be adapted and built on to develop a case-specific survey instrument. In addition, the list of key elements of cooperation and aspects of effectiveness, in Appendices A and B, respectively, can be used to develop context-specific questions. Once the survey instrument is developed, interviews should be conducted with all of the actors identified in the stakeholder analysis in Stage 1. After this data is collected, the user can move on to Stage 3: Evaluation.

Stage 3: Evaluation

The evaluation stage is the final component of the Weighted Model of Effective Cooperation. From the data assessment, the user should have, for each of the Four Frames of Cooperation, baseline/ideal data and current point in time data for each actor. The user should also have data from the standardized questions; these will be compared to either the baseline/ideal data or the results of secondary data research, such as in the example of whether there is a dispute resolution mechanism in a legal arrangement. The evaluation takes place in two steps: (1) comparison and (2) aggregation and synthesis. The data for each related question set between the baseline/ideal and the current point in time should be compared according to one or more of the four methods of evaluation presented earlier in the chapter –goal attainment, problem solving, compliance, and interest-based. Figure 9.7 presents several framing questions for how to compare these data within either of the four methods. Some of the methods are more relevant to data collected within specific frames, which is also exemplified in Figure 9.7. Recall the dispute resolution mechanism example. If there is a dispute resolution mechanism in the legal arrangement, this could be evaluated as compliance with provisions in either of the two global conventions or it could be evaluated as goal attainment if this was advocated for by the actors, or even both, as this evaluation depends on the context. In the Columbia River Treaty example, one of the problems the agreement aimed to solve was future flooding in US cities along the river. The problem was solved by the agreement and the greater cooperative process between the two states. Therefore, we can say that it was effective in problem solving within the outcome frame for this particular problem. However, it was not effective in addressing the standardized goal of including environmental provisions in the cooperative process.

Through the Columbia River example, I am also illustrating that there will be multiple evaluations of the extent of effectiveness within each frame for the various question pairs. This will allow for targeted identification of where a cooperative process may be less effective, and targeted efforts can be made to encourage the development or improvement of effective cooperation in those areas. However, the many evaluations within each frame must then be aggregated into an overall score for each frame. This is a challenging piece of

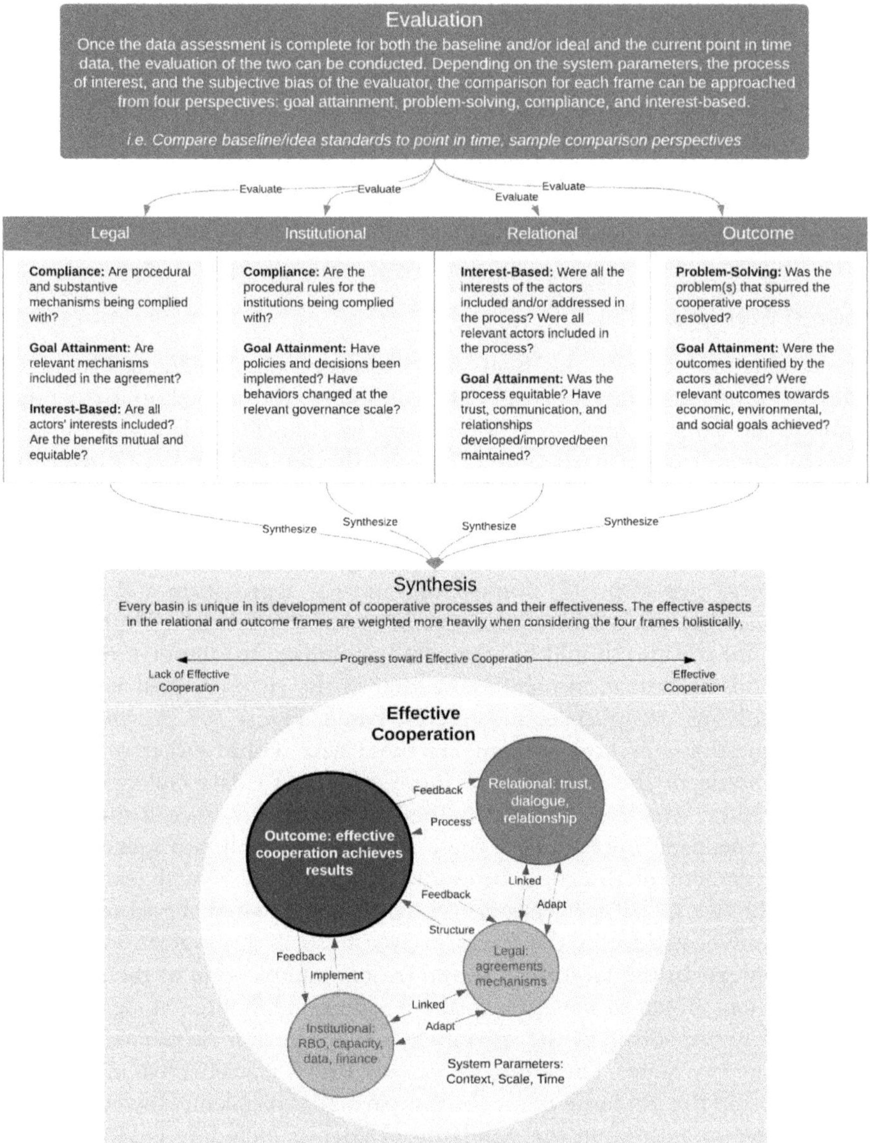

Figure 9.7 Stage 3: Evaluation of the Weighted Model of Effective Cooperation.

the Weighted Model, as in the current stage of development of the model, there is no precise method for aggregating the results since how this is done will depend on the context. For example, having a dispute resolution mechanism present and having a signed treaty are very different, and therefore, the degree that each contributes to the overall effectiveness of the legal frame will vary. Once each frame has been aggregated, we can again aggregate the frames to determine the overall extent of effective cooperation. It is here that the evaluation of the relational and outcome frames should be more heavily emphasized than the legal and institutional frames.

The Weighted Model of Effective Cooperation does not claim to determine causality, nor anything more than an ordinal level of measurement for effectiveness. At its most basic level, the scale can be divided into three categories: lack of effective cooperation, progress in effective cooperation, and effective cooperation. However, depending on how the data are collected, the progress in effective cooperation could be subdivided to show more measurable growth in developing effective cooperation. For example, the Likert scale could be used when ranking trust, communication, or dialogue, allowing for more depth and nuance in the data. The power of this model is less in the aggregated value – lack of effective cooperation or effective cooperation – and more in the detail generated through the frames, the data collection method, and the process of comparison between the baseline/ideal and current point in time. It is here where actors can see where cooperative processes are excelling, weak, or lacking, and they can hopefully encourage more equitability and sustainability in future adaptions and evolutions of the cooperative process. In addition, the results from each actor can be compared to see how equitable a cooperative process truly is. If there are significant differences between actors, then it would contribute to an overall lower evaluation for the cooperative process of interest.

Overall, using the Weighted Model of Effective Cooperation is a complex and intensive process. Each of the three stages requires context-specific information to develop a context-relevant evaluation of the extent of effective cooperation. However, this method addresses some of the critiques of other measurements of cooperation, which are often procedural in nature and focus on more easily measurable elements without evaluating the impact or relationships in the cooperative process. In addition, this method recognizes that cooperation is dependent on the basin or aquifer, the actors involved, the problem at hand, and the other human and physical factors, which means that it is not comparable between basins or aquifers. What is considered effective in the Columbia River Basin may be entirely different than what is considered effective in the Jordan River Basin, and that is just fine. Comparing the same basin over two or more points in time is possible, such as for the purpose of measuring progress; however, the user needs to be careful in recognizing that the context changes over time within the same geographic or governance scale. For example, if we were comparing the effectiveness of the cooperation in the Nile River Basin in the early 1900s with today, there have been considerable changes in the politics and development in the basin. Given the drastic differences between these two points in time, I would argue that they should not be directly compared but instead be compared to several intermediary baselines to truly capture the growth of cooperation in the basin.

Conclusion

This chapter synthesizes the data and discussions on definitions of cooperation and effectiveness presented in the previous Chapters 4 and 7. In addition, it builds on the key elements of cooperation and the aspects of effectiveness explored in Chapters 5 and 8, respectively. Using the Four Frames of Cooperation (see Chapter 6) and the regime effectiveness literature, particularly the work of Underdal, as a foundation, this chapter presents the Weighted Model of Effective Cooperation as the answer to the second challenge posed in Part III: What makes transboundary water cooperation effective and how can we evaluate it? This model can be used conceptually to understand what effective cooperation is in a general sense; see Figures 9.1 and 9.2. It also operationalizes the model in order to actually evaluate the extent of effective cooperation in a basin or aquifer. The complete flowchart for how to use the Weighted Model of Effective Cooperation is in Chapter 10. While the model needs further testing and refinement, it pushes forward the understanding of transboundary water cooperation in two main ways. The first is that relationships – i.e., the relational frame – matter. This is not a controversial point. Nearly every expert I interviewed mentioned relational elements of cooperation, yet this had not been thoroughly addressed in the literature on transboundary water cooperation, particularly in other methods measuring cooperation. The second is how context was addressed. Again, the fact that cooperation and effective cooperation are context-dependent is not disputed. But rather, it is difficult to quantify, measure, and compare cooperation without generalizing context, especially when conducting a global or large-scale analysis between multiple basins or countries, such as SDG Indicator 6.5.2. This can be seen in the regime effectiveness literature, where much of the research is on specific case studies that allow context to play a role. My goal with the Weighted Model is to enable context to be taken into account when evaluating cooperation while also providing a reasonably similar model for any basin or aquifer at various scales.

Notes

1 Please keep in mind when reading that this chapter focuses on defining *transboundary water cooperation*, not cooperation in the broader sense of the word. I use *cooperation* without the *transboundary water* modifier; however, I am referring to *transboundary water cooperation*. I will note when I refer to cooperation unlinked to transboundary water.
2 This outcome is defined as the changes in human behavior as a result of the regime. This differs from the definition of the outcomes and outcome frame that are used within the Four Frames of Cooperation, which are the benefits, goals, actions, or specific outputs of cooperation.

References

Benson, David, and Andrew Jordan. 2015. "Environmental Policy: Protection and Regulation." In *International Encyclopedia of the Social & Behavioral Sciences (Second*

Edition), edited by James D. Wright, 778–83. Oxford: Elsevier. https://doi.org/10.1016/B978-0-08-097086-8.91014-6.

Bernauer, Thomas. 1995. "The Effect of International Environmental Institutions: How We Might Learn More." *International Organization* 49 (2): 351–77. https://doi.org/10.1017/S0020818300028423.

Boklan, Daria S., and Barbara Janusz-Pawletta. 2017. "Legal Challenges to the Management of Transboundary Watercourses in Central Asia under the Conditions of Eurasian Economic Integration." *Environmental Earth Sciences* 76 (12). https://doi.org/10.1007/s12665-017-6741-3.

Bourne, Charles B. 1998. "The Primacy of the Principle of Equitable Utilization in the 1997 Watercourses Convention." *Canadian Yearbook of International Law / Annuaire Canadien de Droit International* 35: 215–32. https://doi.org/10.1017/S0069005800006639.

Brundtland Commission. 1987. "Our Common Future." World Commission on Environment and Development. https://sustainabledevelopment.un.org/content/documents/5987our-common-future.pdf.

Conca, Ken, Fengshi Wu, and Cigi Mei. 2006. "Global Regime Formation or Complex Institution Building? The Principled Content of International River Agreements." *International Studies Quarterly* 50 (2): 263–85. https://doi.org/10.1111/j.1468-2478.2006.00402.x.

Daoudy, Marwa, and Elizabeth J. Kistin. 2008. *Beyond Water Conflict: Evaluating the Effects of International Water Cooperation.* At the 49th Annual Conference of the International Studies Association on 26–29 March 2008 held in San Francisco. San Francisco.

Devlaeminck, David. 2018. Research Interview.

Dombrowsky, Ines. 2007. *Conflict, Cooperation and Institutions in International Water Management: An Economic Analysis*. Cheltenham, United Kingdom: Edwards Elgar Publishing.

Eckstein, Gabriel. 2018. Research Interview. Skype.

Euston, Stanley R., and William E. Gibson. 1995. "The Ethic of Sustainability." *Earth Ethics* 6, 5–7.

Graddy, Garrett. 2011. "Politics of Scale." In *Green Politics: An A-to-Z Guide*, edited by Dustin Mulvaney, 328–31. Los Angeles: Sage Publications.

Grey, David, Claudia Sadoff, and Genevieve Connors. 2009. "Effective Cooperation on Transboundary Waters: A Practical Perspective." In *Getting Transboundary Water Right Theory and Practice for Effective Cooperation*, 15–20. Stockholm: SIWI.

Helm, Carsten, and Detlef Sprinz. 2000. "Measuring the Effectiveness of International Environmental Regimes." *Journal of Conflict Resolution* 44 (5): 630–52. https://doi.org/10.1177/0022002700044005004.

Huntjens, Patrick, Yumiko Yasuda, Ashok Swain, Rens de Man, Bjørn-Oliver Magsig, and Shafiqul Islam. 2016. *The Multi-Track Water Diplomacy Framework: A Legal and Political Economy Analysis for Advancing Cooperation over Shared Waters*. Hague: The Hague Institute for Global Justice. www.thehagueinstituteforglobaljustice.org/thinking-doing/publication/the-multi-track-water-diplomacy-framework-a-legal-and-political-economy-analysis-for-advancing-cooperation-over-shared-waters/.

Kistin, Elizabeth J., and David .J.H. Phillips. 2007. "A Critique of Existing Agreements on Transboundary Waters and Proposals for Creating Effective Cooperation between Co-Riparians." *Third International Workshop on Hydro-Hegemony*. London.

Lebel, Louis, Po Garden, and Masao Imamura. 2005. "The Politics of Scale, Position, and Place in the Governance of Water Resources in the Mekong Region." *Ecology and Society* 10 (2): 18.

Miles, Edward, Arild Underdal, Steinar Andresen, Jørgen Wettestad, Jon Birger Skjaerseth, and Elaine Carlin, eds. 2002. *Environmental Regime Effectiveness: Confronting Theory with Evidence*. Cambridge, MA: MIT Press.

Mirumachi, Naho. 2018. Research Interview.

Purvis, Ben, Yong Mao, and Darren Robinson. 2019. "Three Pillars of Sustainability: In Search of Conceptual Origins." *Sustainability Science* 14 (3): 681–95. https://doi.org/10.1007/s11625-018-0627-5.

Rafaelli, Silvia. 2018. Research Interview.

Schmeier, Susanne. 2013. *Governing International Watercourses: River Basin Organizations and the Sustainable Governance of Internationally Shared Rivers and Lakes*. New York: Routledge. www.routledge.com/Governing-International-Watercourses-River-Basin-Organizations-and-the/Schmeier/p/book/9780415623582.

———. 2018. Research Interview.

Sindico, Francesco. 2018. Research Interview.

Sugg, Zachary P. 2018. Research Interview.

Svendsen, Mark (Ed). 2005. *Irrigation and River Basin Management: Options for Governance and Institutions*. CABI; International Water Management Institute (IWMI). https://cgspace.cgiar.org/handle/10568/36480.

Underdal, Arild. 1992. "The Concept of Regime 'Effectiveness.'" *Cooperation and Conflict* 27 (3): 227–40.

———. 2002. "One Question, Two Answers." In *Environmental Regime Effectiveness: Confronting Theory with Evidence*, edited by Edward Miles, Arild Underdal, Steinar Andresen, Jørgen Wettestad, Jon Birger Skjaerseth, and Elaine Carlin, 3–43. Cambridge, MA: MIT Press.

United Nations. 1997. *Convention on the Law of Non-Navigational Uses of International Watercourses. A/RES/51/229.* http://heinonline.org/hol-cgi-bin/get_pdf.cgi?handle=hein.journals/tilj33&section=22.

Wolf, Aaron T. 2010. *Sharing Water, Sharing Benefits: Working towards Effective Transboundary Water Resources Management. A Graduate/Professional Skills-Building Workbook.* UNESCO: International Hydrological Programme: Oregon State University. http://unesdoc.unesco.org/images/0018/001893/189394e.pdf.

Wolf, P.J. 2001. "Authority: Delegation." In *International Encyclopedia of the Social & Behavioral Sciences*, edited by Neil J. Smelser and Paul B. Baltes, 972–78. Oxford: Pergamon. https://doi.org/10.1016/B0-08-043076-7/04476-4.

Young, Oran R. 1994. "The Effectiveness of International Governance Systems." In *International Governance: Protecting the Environment in a Stateless Society*, by Oran R. Young, 140–60. Ithaca: Cornell University Press.

———. 2001. "Inferences and Indices: Evaluating the Effectiveness of International Environmental Regimes." *Global Environmental Politics* 1 (1): 99–121. https://doi.org/10.1162/152638001570651.

———. 2011. "Effectiveness of International Environmental Regimes: Existing Knowledge, Cutting-Edge Themes, and Research Strategies." *Proceedings of the National Academy of Sciences* 108 (50): 19853–60.

Young, Oran R., and Marc Levy. 1999. "The Effectiveness of International Environmental Regimes." In *The Effectiveness of International Environmental Regimes: Causal Connections and Behavioral Mechanisms*, edited by Oran R. Young, 1–32. Cambridge, MA: The MIT Press.

Zawahri, Neda A. 2008. "Designing River Commissions to Implement Treaties and Manage Water Disputes: The Story of the Joint Water Committee and Permanent

Indus Commission." *Water International* 33 (4): 464–74. https://doi.org/10.1080/02508060802474566.

Zeitoun, Mark. 2018. Research Interview.

Zeitoun, Mark, Ana Elisa Cascão, Jeroen Warner , Naho Mirumachi, Nathanial Matthews, Filippo Menga, and Rebecca Farnum. 2017. "Transboundary Water Interaction III: Contest and Compliance." *International Environmental Agreements: Politics, Law and Economics* 17 (2): 271–94. https://doi.org/10.1007/s10784-016-9325-x.

Zeitoun, Mark, and Naho Mirumachi. 2008. "Transboundary Water Interaction I: Reconsidering Conflict and Cooperation." *International Environmental Agreements: Politics, Law and Economics* 8 (4): 297–316. https://doi.org/10.1007/s10784-008-9083-5.

Zeitoun, Mark, Naho Mirumachi, and Jeroen Warner. 2011. "Transboundary Water Interaction II: The Influence of 'Soft' Power." *International Environmental Agreements: Politics, Law and Economics* 11 (2): 159–78. https://doi.org/10.1007/s10784-010-9134-6.

Ziganshina, Dinara. 2018. Research Interview.

10 A path forward for effective cooperation over transboundary waters

Over four billion people reside in a transboundary river basin; this number is even higher when we include transboundary aquifers (TFDD 2021). Not only are humanity's domestic, industrial, and agricultural water needs dependent on transboundary water sources, vital ecosystems and diverse habitats are also dependent on these resources. In order to share these waters in the face of climate change, rising demand, increased threats to quality, and other challenges, cooperation between states, sectors, and users is more important than ever. However, as discussed in this book and through the work of other scholars, cooperation is not a catch-all solution or an always positive process for shared users. Therefore, to ensure that cooperative processes yield sustainable and equitable sharing of transboundary resources, we need to understand better what makes cooperation effective. By addressing this point, we can develop and shape transboundary cooperative processes to be more effective at addressing our global waters' coming challenges.

The need to better understand what makes cooperation effective can be broken down into two main components, as identified and demonstrated in Part I (Chapters 2 and 3). The first challenge is that there is no single definition for transboundary water cooperation (Tarlock 2015; Saruchera and Lautze 2015). This presents difficulties when trying to establish and measure cooperation[1] consistently and comparably. This challenge has led scholars to argue for measuring cooperation based on the outputs of cooperation – that cooperation should be meaningful, have measurable benefits, and, ultimately, be effective. This leads us to the second challenge of defining effectiveness in the context of transboundary water cooperation. Effectiveness is often not defined explicitly and is subjective, which leads to multiple interpretations of the term. In their work on effective cooperation, Daoudy and Kistin (2008, 23) recognize this stating "So while there is no single strategy, clarity is needed in the definitions of success that are used and the basis for the methods that are chosen." This book aims to address both of these challenges and, ultimately, provide a tool for evaluating the effectiveness of cooperation over transboundary waters that encourages an increase in the sustainability and equitability of cooperative processes.

DOI: 10.4324/9781003126249-13

Challenge 1: A framework for defining transboundary water cooperation

When I initially began this research, I assumed defining transboundary water cooperation would be a straightforward component – a steppingstone, if you will, to the "harder" challenge of evaluating effectiveness. However, through data collection and analysis, I realized that this was a false assumption. Instead of a steppingstone to exploring effectiveness in the context of transboundary water cooperation, identifying the core elements and developing a clear definition of transboundary water cooperation were the foundation of this process. Without these elements and a clear definition, any analysis of effective cooperation is limited.

Most definitions in the literature are implicit, meaning that cooperation is rarely defined, but rather it is assumed that the reader knows what is meant by the term 'transboundary water cooperation.' Occasionally, transboundary water cooperation is defined explicitly in the literature, with definitions falling into five categories: definitions of general cooperation, examples of cooperation, frameworks of cooperation, antithesis definitions, and direct, explicit definitions. There were only 23 direct, explicit definitions in the literature sample of 273 articles. These data support the claim that there is no single, accepted definition of transboundary water cooperation in the field, nor of general cooperation (Phillips et al. 2006; Woodhouse and Zeitoun 2008; Tarlock 2015). This lack of a concise and clear definition that is agreed upon in the field can create complications when accounting for the resolution of water disputes (Postel and Wolf 2001; Gerlak and Zawahri 2009).

Fifty key elements of cooperation are identified as standalone concepts commonly referenced in implicit and explicit discussions of cooperation (see Appendix A), of which treaty or agreement, mutual benefit, institutional capacity, data and information, and basin planning are the top five most commonly referenced. However, a complicating factor is the frequent use of water governance, diplomacy, and, to a lesser extent, management used interchangeably with transboundary water cooperation. While these terms overlap, the confusion between them stems from the cooperation with a 'lowercase c' that is inherent in each of the terms. Cooperation with a 'lowercase c' is a joint action that can be done cooperatively or non-cooperatively, while transboundary water cooperation is defined more specifically as the cooperative process that embodies the key elements identified. A concise definition is derived from these key elements, where transboundary water cooperation is the *interactions between actors over shared waters that result in establishing mutually beneficial outcomes through a decision-making process; this process could include formal and informal legal and institutional mechanisms depending on the scale and context.* While comprehensive of the main themes used to describe and measure cooperation, it has the same limitations as other statement-style definitions. To be concise and applicable to many situations, it is also broad and non-specific, making it hard for the definition to indeed be useful in defining what is or is not cooperation in a particular basin or aquifer.

Therefore, to address the first challenge, this book proposes a framework definition of transboundary water cooperation that can be practically applied and gives the flexibility to accommodate different contexts, scales, and places. The Four Frames of Transboundary Water Cooperation is a framework for developing an explanatory definition of transboundary water cooperation for a particular basin or aquifer at the relevant scale. The key elements of cooperation fall into four frames – legal, institutional, relational, and outcome. These four frames capture the variability in what key elements are applicable and relevant to a particular cooperative process. Figure 10.1 summarizes the Four Frames of Cooperation and includes definitions of each of the four frames.

An important aspect to keep in mind when using the framework is that this is a neutral platform for defining cooperation in a context-specific yet consistent manner. It is not passing a value judgment on the cooperative process; instead, it provides a means to comprehensively identify the components and elements that compose a cooperative process. However, like other definitions of transboundary water cooperation, this does not determine whether cooperation

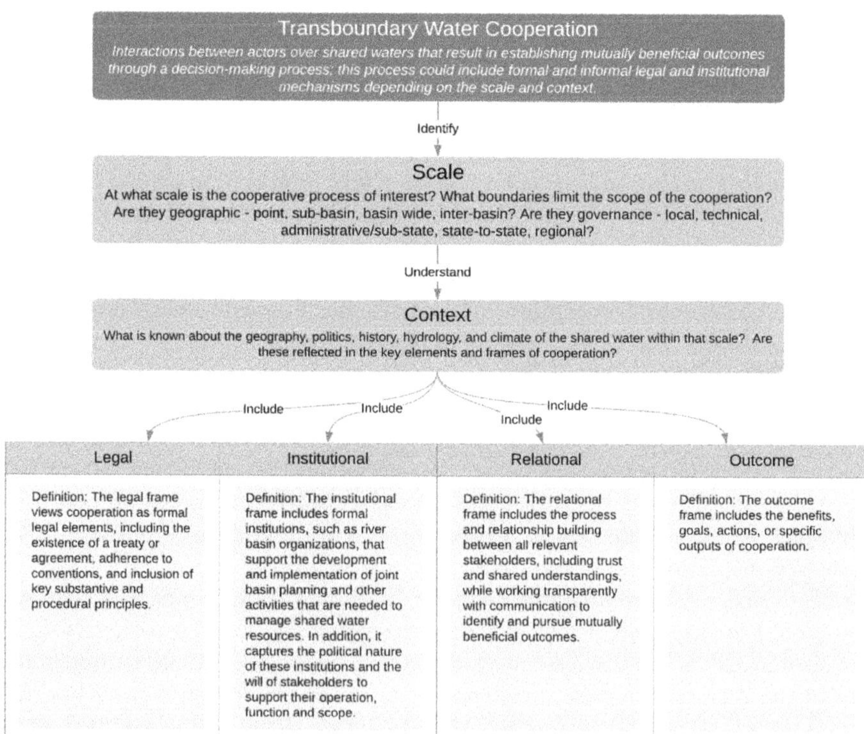

Figure 10.1 Summary of the Four Frames of Cooperation.

is meaningful or whether it will produce measurable benefits (Tarlock 2015; Strategic Foresight Group 2015). To evaluate the quality of cooperation, we must address the second challenge to evaluate whether cooperation is effective or not.

Challenge 2: A method for evaluating effective transboundary water cooperation

If we aim to evaluate the effectiveness of transboundary water cooperation, we must also define what is meant by effectiveness. As with cooperation, effectiveness is not frequently defined in the literature, either implicitly or explicitly. There were only five explicit definitions of effective transboundary water cooperation in the literature. In many cases, defining or describing what effective cooperation *was not* seemed to be quite common in the data.

Within the implicit and explicit discussions of effective cooperation, there are 49 aspects of effectiveness (see Appendix B), with the top five most referenced aspects of effectiveness being: achieve outcomes, participation, mutual satisfaction, multi-scale, and define outcomes. In addition to the aspects of effectiveness, barriers to effective cooperation are also identified. Overall, the top most commonly referenced barriers to effective cooperation include context, political will, national level, financing, and lack of knowledge. Using the aspects of effectiveness, barriers, and the foundational definition developed for transboundary water cooperation, a concise definition is established: Effective transboundary water cooperation *achieves equitable results or outcomes through a cooperative decision-making process that has trusting relationships with stakeholder participation and communication. This process is supported by the appropriate legal and institutional arrangements, where the outcomes can be the resolution of the initial problems, reaching the goals of the parties, or complying with mutually agreed-upon principles in a legal arrangement.* Like the concise definition of cooperation, this definition is general and broad, and it cannot be operationalized to measure whether cooperation is effective or not. Instead, it presents a more general summary of the main aspects of effective cooperation. So, the question remains – how is effective cooperation evaluated?

The aspects of effectiveness, as well as the barriers to effective cooperation, fit within the Four Frames of Cooperation, which allows us to begin developing a structure for what effective cooperation looks like in transboundary waters. This structure is based on this book's main findings: the relational and outcome frames of cooperation are perceived as being more relevant to effective cooperation than the legal and institutional frames. While not counterintuitive, this differs from most of the transboundary water cooperation measures found in the international waters literature. For example, transboundary water cooperation is commonly measured by the presence of an agreement or the inclusion of specific provisions in an agreement. Relationships, trust, communication, and dialogue between stakeholders, as well as achieving the outcomes agreed upon between parties, are more commonly perceived as effective cooperation in a basin than a signed agreement.

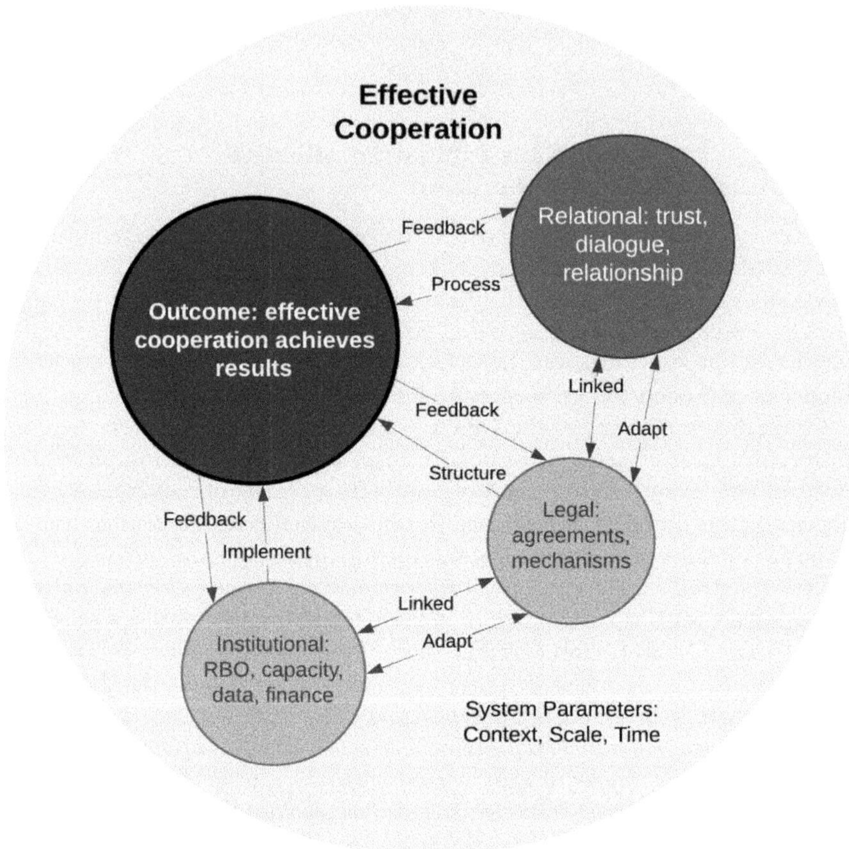

Figure 10.2 The Weighted Model of Effective Cooperation.

This finding – that relation and outcome frames are perceived as most important for effective cooperation – is incorporated into the Weighted Model of Effective Transboundary Water Cooperation. This model forms the basis for evaluating effective cooperation, where the aspects in the relational and outcome frames are weighted more heavily in the evaluation. The Weighted Model is depicted graphically in Figure 10.2, which shows the outcome and relational frames as larger and darker in color than the legal and institutional frames. This is not done to negate the importance of the legal and institutional elements but rather to illustrate that fully effective cooperation is not possible without progress toward achieving outcomes or developing trusting, transparent relationships between actors.

The final step in addressing the second challenge is operationalizing the Weighted Model so that it can evaluate whether a cooperative process is effective. Chapter 9 lays out a three-stage process for using the Weighed Model (see Figure 10.3). In order to conduct the evaluation, the user must first define the process of interest – what problem, issue, or cooperative process is being evaluated. This can then be refined by considering the other parameters that will affect the evaluation, including scale, context, and time. The scale of meaning needs to be identified, which is the scale at which the process or problem is discussed or analyzed. This may differ from the scale at which the cooperative process should occur according to the principle of subsidiarity. In addition, the point(s) in time that the cooperative process is being evaluated at should be identified. These parameters will influence the contextual factors that are influencing and informing the formation, operation, and impact of the cooperative process.

The cooperative process is not the only thing that needs to be defined according to the parameters. The standards that it will be compared to in order to determine its effectiveness also need to be defined. The unit of measurement for this model is that *effective cooperation is equitable and sustainable*. This is the goal cooperation should be working toward to be considered effective – how this will appear and mechanisms that achieve this will vary depending on the context.

The other standard is the point of reference for comparison. The comparison can be done in two ways: by comparing to a baseline or to a hypothetical ideal. By comparing to the baseline, or multiple baselines, we can track progress over time. By comparing to the hypothetical ideal, we can determine the extent to which the process is considered effective and areas where efforts to improve effectiveness can be concentrated. These standards inform the development of the assessment instrument. Along with the parameters, the standards help determine what questions should be asked to all of the relevant actors to the process of interest in both the baseline/ideal and the current point in time for comparison.

The evaluation is the final stage in evaluating effective cooperation using the Weighted Model. This stage provides four methods to compare the two sets of data from the baseline/ideal and the current point in time: compliance, goal attainment, problem solving, and interest-based. The model does not determine causality in these comparisons; these are merely a means to evaluate and track progress as perceived by the actors. Furthermore, comparison based on specific key elements or aspects of effectiveness provides disaggregated data on effectiveness. Users can then identify areas where cooperation is ineffective. By aggregating to the level of each of the four frames, the model makes overall discussions easier on which frames are the most effective or not for a cooperative process. Whether a key element, frame, or the process overall is effective or not is limited to the ordinal scale of measurement. The Weighted Model proposes three levels – lacking effective cooperation, progress toward effective cooperation, and effective cooperation. Given the subjective nature of the questions used to collect information on the process and the subjectivity in the evaluation, a more rigorous breakdown is not possible at this point. When

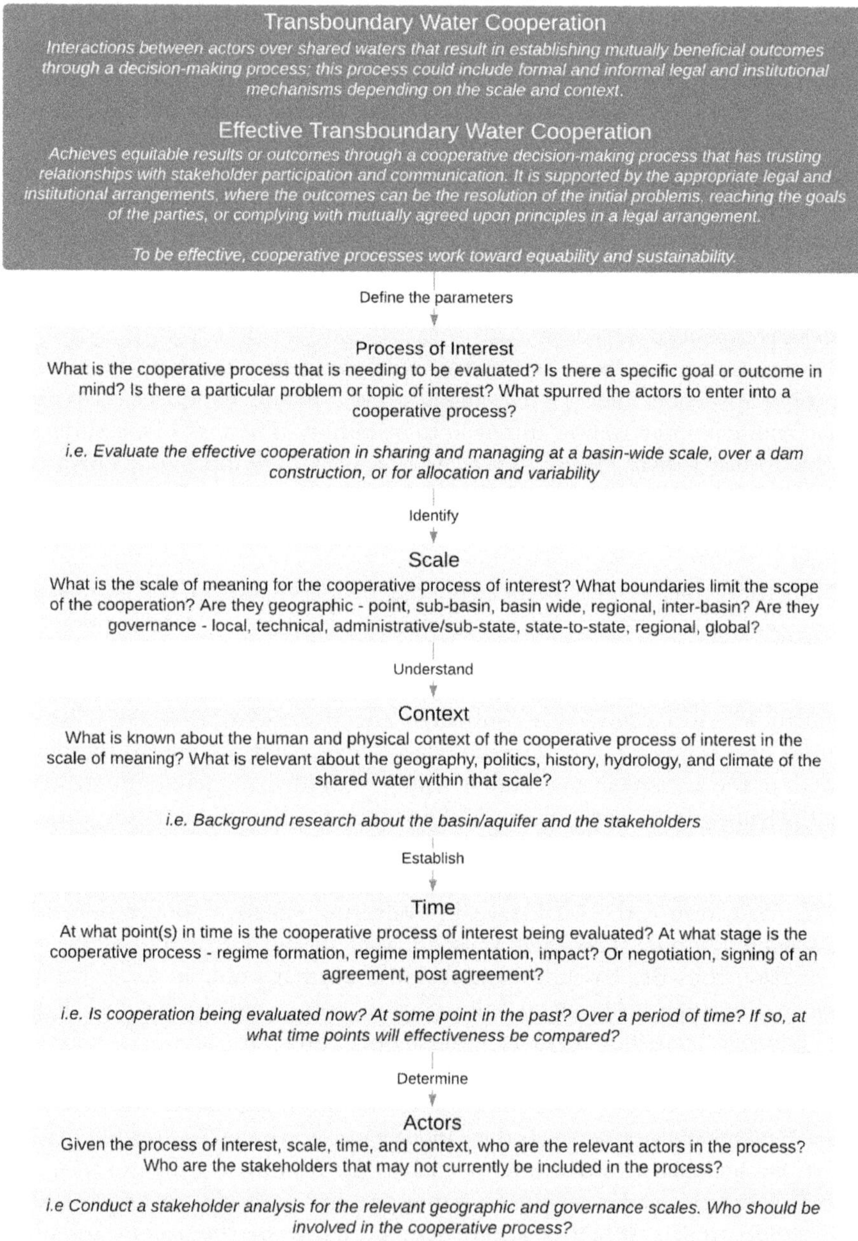

Figure 10.3 Complete flow chart for evaluating the effectiveness of transboundary water cooperation using the Weighted Model of Effective Transboundary Water Cooperation.

Figure 10.3 Continued

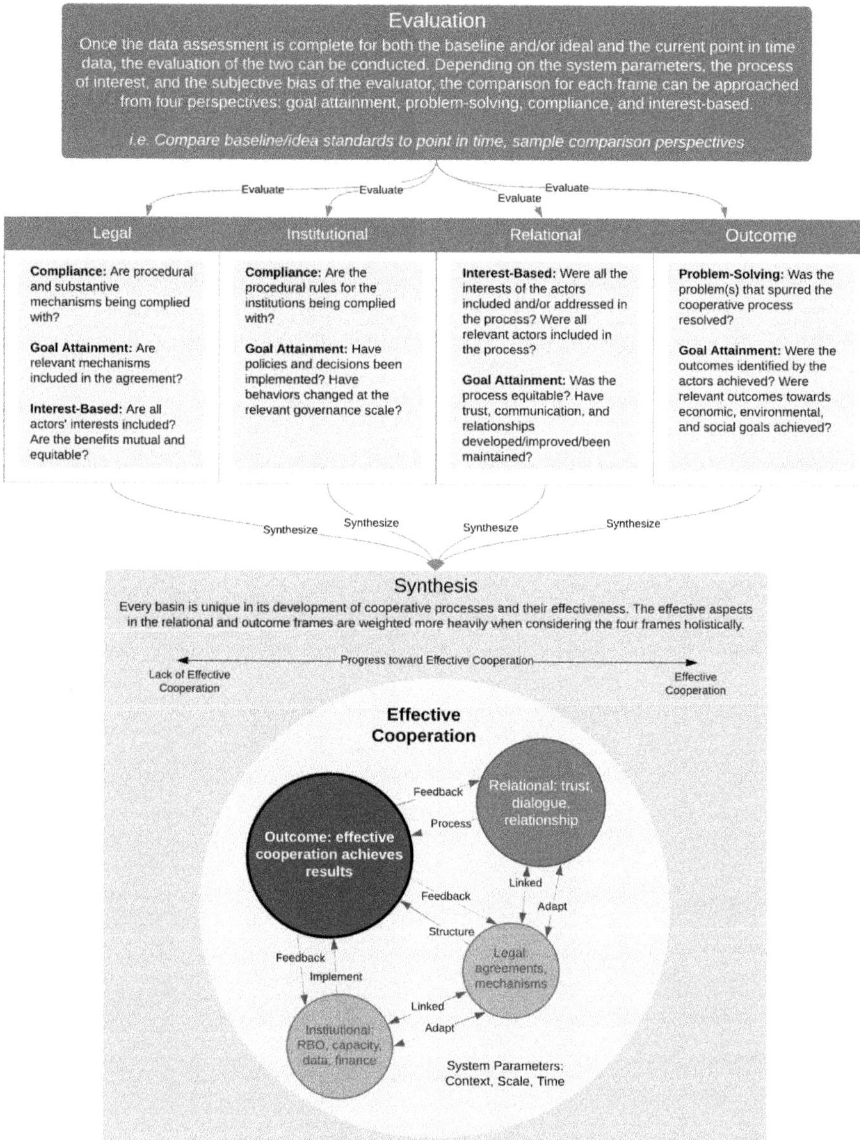

Figure 10.3 Continued

aggregating to an overall process score, the level of effectiveness in the relational and outcome frames needs to be weighted more heavily for their contribution to the overall effectiveness in the basin or aquifer, as illustrated in Figures 10.2 and 10.3, which graphically depicts the Weighted Model and the flow chart for operationalizing the model, respectively.

Considerations for application

While my hope is that these frameworks will be helpful to academics, practitioners, and policymakers alike, they are not without critique. The predominant point of consideration is that causality cannot be drawn between cooperation and key elements of cooperation nor effectiveness and the aspects of effectiveness. Given the study's design, the complexity of the topics, and the variability between basins and aquifers, there is no means to determine which key elements or aspects actually result in improved or effective cooperation. Instead, the data illustrate the key elements or aspects that are perceived as relevant to cooperation and effectiveness. While the key elements and aspects were evaluated based on the frequency of reference in the literature and interview data, this does not correlate to a ranking of the key elements and aspects relative to cooperation and effectiveness. The most heavily referenced key elements and aspects are perceived as the most applicable across various basin and aquifer contexts, more than they are perceived as important to have in a basin or aquifer. Importance will depend on the context and the issue requiring cooperation.

With respect to applying the Weighted Model to evaluate effective cooperation, the user should note that it does not determine causality when evaluating effective cooperation. The user cannot draw a link from one of the key elements or aspects of effectiveness and identify it as a cause of a cooperative process achieving or not achieving a specific outcome. Rather, the model provides a means to consider *all* of the potential key elements and aspects that *could* influence effective cooperation. The ability to draw causal links is an area where future research is needed, potentially using historical case studies where the outcomes and effects of cooperation have already been achieved or abandoned, and causal relationships can be drawn.

Another factor for consideration when using the Weighted Model is the inherent subjectivity in the evaluation of effectiveness. While the model attempts to reduce the subjectivity by defining the standard unit of measure as equitable and sustainable, how that unit is interpreted in a particular basin, aquifer, or for a specific cooperative process may differ. For example, in two different fossil aquifers, one set of actors may agree that sustainable use of the groundwater is controlled extraction over the next few decades until it has been fully consumed. In contrast, the actors in the other fossil aquifer may define sustainable as agreeing to not extract the groundwater, except under certain pre-agreed circumstances. This case also illustrates the final consideration of the model – comparability. Since cooperation is case-specific and effectiveness is subjective, the extent of effectiveness as evaluated by the Weighted Model is not comparable between different basins and aquifers. By comparing the basins or aquifers, the nuance and detail that is embedded in the model have the potential to be overlooked, and the context that the cooperative process operates within may be minimized. Furthermore, the questions in the assessment instrument and the methods of evaluation will differ, making a comparison between basins and aquifers even more inadvisable.

A path forward

This book addresses both the challenges identified in the international waters field related to transboundary water cooperation. It develops concise statement definitions for both transboundary water cooperation and effective transboundary water cooperation:

- **Transboundary water cooperation**: interactions between actors over shared waters that result in establishing mutually beneficial outcomes through a decision-making process; this process could include formal and informal legal and institutional mechanisms depending on the scale and context.
- **Effective transboundary water cooperation**: achieves equitable results or outcomes through a cooperative decision-making process that has trusting relationships with stakeholder participation and communication. This process is supported by the appropriate legal and institutional arrangements, where the outcomes can be the resolution of the initial problems, reaching the goals of the parties, or complying with mutually agreed-upon principles in a legal arrangement. Effective cooperation is equitable and sustainable.

However, as discussed, these definitions have limitations, given that both terms are context-specific and unique for every basin or aquifer. Therefore, this book pushes past concise definitions to use frameworks and models for defining transboundary water cooperation and evaluating effectiveness while recognizing the inherent variability in terminology, as well as the inherent subjectivity in defining effectiveness. The two new frameworks – the Four Frames of Cooperation and the Weighted Model of Effective Cooperation – improve our conceptualization of transboundary water cooperation by evaluating the quality of that cooperation and considering whether it has been effective, based on the principles of equitability and sustainability. They encourage detailed evaluation of the quality, success, or effectiveness of cooperative processes.

So, how do we move forward from here? There are two directions for application and then for refinement and expansion. The Four Frames of Cooperation and the Weighted Model of Effective Cooperation can be applied by users who want to understand cooperation in a transboundary water context better.

Furthermore, the power of the Weighted Model and its assessment process is in the disaggregated data and context-specific assessment, which can support users in developing cooperative processes that are effective in meeting future challenges by being equitable and sustainable.

Alongside the application of the framework and the model, further refinement and expansion of the methodology are needed. In particular, the Weighted Model and the methodology for evaluating effective transboundary water cooperation needs to be refined and validated. This can be done through the comparison to other methods proposed for measuring transboundary water cooperation, such as the Water Cooperation Quotient or the Hydropolitical Tension indicator in the TWAP-RB, and to other methods that measure effectiveness, such as a treaty, RBO, or institutional effectiveness. It would be interesting to compare what aspects of cooperation and effectiveness overlap and differ between the methods used in the international waters field, as well as other methods for assessing cooperation on other transboundary natural resources.

The methodology has the potential to be expanded to better understand effective cooperation within international waters and at other scales and other resources. Further investigation and testing could help to establish causality between key elements of cooperation and aspects of effectiveness with whether a cooperative process is found to be effective. Lastly, the Weighted Model could have the potential to be transferred to other scales and other sectors. Future research should explore how the model could be adapted to non-international transboundary waters, such as intrastate waters or even shared water between groups in the same community. Similarly, how the Weighted Model could be used in other transboundary environmental sectors where a natural resource is being shared.

In Chapter 1, I stated that this book is about definitions at its core, but it is much broader in scope in practicality. I hope that through reading, you can take away a new perspective on transboundary water cooperation and its effectiveness, particularly related to their complexity, nuance, and context. In addition, I hope you take away three other main points: (1) the importance of place, scale, and context in the development and continuation of cooperative processes. The relational and outcome frames are influenced by these factors and are central to effective cooperation; (2) that cooperation is not always positive, and reframing transboundary water cooperation as a neutral process with the potential to contribute to both positive and negative outcomes will enable more effective processes; and (3) that for cooperation to be effective, it should be equitable and sustainable. As UN Secretary General Ban Ki-Moon expressed, there is no other option but international cooperation over water. "Water is a classic common property resource. No one really owns the problem. Therefore, no one really owns the solution" (UN News 2008). Nevertheless, I would add that while no one actor owns the solution to these collective action problems, we are all responsible for ensuring that our interactions over transboundary waters are sustainable and equitable.

Note

1 Please keep in mind that this chapter and the previous chapters focus on *effective transboundary water cooperation*. Therefore, I use *effective cooperation* and *cooperation*, without the 'transboundary water' modifier, to still mean *effective transboundary water cooperation* or *transboundary water cooperation*, respectively. I will note if I am referring to a broader sense of either effective/effectiveness or cooperation that is unlinked from transboundary water.

References

Daoudy, Marwa, and Elizabeth J. Kistin. 2008. *Beyond Water Conflict: Evaluating the Effects of International Water Cooperation*. At the 49th Annual Conference of the International Studies Association on 26–29 March 2008 in San Francisco.

Gerlak, Andrea, and Neda Zawahri. 2009. "Navigating International River Disputes to Avert Conflict." *International Negotiation* 14 (2): 211–27. https://doi.org/10.1163/157180609X432806.

Phillips, David, Marwa Daoudy, Stephen McCaffrey, Joakim Öjendal, and Anthony Turton. 2006. *Trans-Boundary Water Cooperation as a Tool for Conflict Prevention and Broader Benefit-Sharing*. Ministry for Foreign Affairs Stockholm.

Postel, Sandra L., and Aaron T. Wolf. 2001. "Dehydrating Conflict." *Foreign Policy*, 126 (September): 60. https://doi.org/10.2307/3183260.

Saruchera, Davison, and Jonathan Lautze. 2015. "Measuring Transboundary Water Cooperation: Learning from the Past to Inform the Sustainable Development Goals." International Water Management Institute (IWMI). www.iwmi.cgiar.org/publications/iwmi-working-papers/iwmi-working-paper-168/.

Strategic Foresight Group. 2015. "Water Co-Operation Quotient." 2015. http://strategicforesight.com/water-cooperation-map/.

Tarlock, Dan. 2015. "Promoting Effective Water Management Cooperation Among Riparian Nations." *TEC Background Papers*, no. 21. www.gwp.org/Global/ToolBox/Publications/Background%20papers/GWP_TEC21_web.pdf.

TFDD. 2021. "Transboundary Freshwater Dispute Database." College of Earth, Ocean and Atmospheric Sciences, Oregon State University. 2021. https://transboundarywaters.science.oregonstate.edu/.

UN News. 2008. "Ban Ki-Moon Warns That Water Shortages Are Increasingly Driving Conflicts." *UN News*, February 6, 2008. https://news.un.org/en/story/2008/02/248092-ban-ki-moon-warns-water-shortages-are-increasingly-driving-conflicts.

Woodhouse, Melvin, and Mark Zeitoun. 2008. "Hydro-Hegemony and International Water Law: Grappling with the Gaps of Power and Law." *Water Policy* 10 (S2): 103–19. https://doi.org/10.2166/wp.2008.209.

Appendix A: Key elements of cooperation

The key elements of cooperation in Table A.1 identified through the qualitative content analysis are listed in order of most commonly referenced using both the literature sample and the interview responses to the question defining transboundary water cooperation. The description provides a brief summary of the text that is captured in each key element. Text can be coded under multiple key elements if it captures more than one idea.

Table A.1 Key elements of cooperation

Key elements of cooperation	*Codebook description*
Treaty or agreement	Mention a treaty, agreement, or formal legal document signed.
Mutual benefit	Mentions the inclusion of mutual benefits, mutual satisfaction, or similar.
Institutional capacity	References institutional capacity in the exact phrase, or combination of both institution and agreement.
Data and information	Includes data and information exchange, shared data, joint data collection and projects, etc.
Basin planning	Mentions having a basin plan, joint management planning, joint objectives, basin-wide objectives, etc.
Shared understandings	Mentions how parties have common goals, shared interests, acknowledge other parties' interests, etc.
Stakeholder participation	Mentions the inclusion of all relevant stakeholders in the process, equal inclusion of stakeholders in the process.
Communication	Mentions communication, dialogue, and open pathways for discussion.
Trust	References trust. This is in any context, e.g., trust in data, trust in communication, trusting stakeholders, trust in stakeholders, trust in the process.
Equitable	References equitable, equity, equal, etc. This is in any context, e.g., equitable and reasonable, equitable benefits.

(*continued*)

Table A.1 Cont.

Key elements of cooperation	*Codebook description*
Contextual	References aspects of context. This is defined broadly; it includes politics, geography, climate, history, etc.
Process	References the process of cooperation. Often discussed in a more abstract sense of process, rather than identification of specific process elements.
Procedural	Includes text that mentions procedural elements in general or with multiple examples.
Ecological and environmental	References environmental or ecological change, outcomes, or benefits of cooperation, either specifically or generally.
River basin organization	Mentions a river basin organization or similar.
Multiple scales	Includes text that addresses cooperation at multiple scales, specifically or generally.
Technical cooperation	Mentions technical cooperation specifically or mentions cooperation over a technical process or between technical or engineering level people.
Political will	Specifically mentions political will or a variant of the term.
Dispute resolution	Mentions the inclusion of a dispute resolution mechanism in a legal agreement or within an institution.
Sustainable	Mentions sustainable, or its variants, generally with respect to outcomes of cooperation.
Relationship building	Mentions relationships between actors, with respect to building, developing, maintaining, trusting, long term, etc. Could be at multiple scales.
Human capacity	Mentions human capacity specifically, or the technical skill, negotiation skill, etc. of people or actors in the process.
Diplomatic cooperation	Mentions the term diplomatic cooperation, or cooperation that is specifically occurring at the diplomatic level, could be defined in a legal agreement, through an institution, informal pathways for diplomatic communication.
Monitoring	Mentions monitoring of data or the water resource. It could be joint or individual.
National action	Mentions national laws, policies, or actions that influence the cooperation or need to be established because of cooperation.
Implementation	Includes text that discusses the implementation of any aspect of cooperation.
Outputs	Mentions outputs, outcomes, benefits, results, etc. It includes both the abstract terms and if any specific outputs are mentioned (these would be coded twice).
Customary law	Mentions customary law in general or lists examples of many customary laws.

Table A.1 Cont.

Key elements of cooperation	*Codebook description*
Substantive elements	Includes text that mentions substantive elements in general or with multiple examples.
Multi-sectoral	Mentions cooperation occurring across several environmental or economic sectors, includes references to integration and IWRM.
Cost-benefit	References cost-benefits of cooperative processes.
Financing	References financing/funding of the cooperative process includes ideas of sustainable financing, actor-based financing, and donor financing.
Cooperation continuum	References cooperation as a continuum, often with particular reference to Sadoff and Grey (2005).
Adaptation	References adaption of the cooperative process, adaptive management, or changing the outcomes, process, benefits, etc. because of changing conditions or new needs or interests.
Utilization/ development	Mentions use of water resources, development of water resources, economic development, and similar as a needed outcome or benefit of the cooperative process.
Transparency	Mentions transparency or a variant, anywhere within the discussion of cooperation.
Allocation	Mentions that there is or needs to have a specific mechanism for allocation, generally within a treaty or agreement.
Stability	Mentions that a cooperative process should be stable, provide stability in the relationship, and other variants.
Informal	Mentions informal aspects of cooperation or that informal cooperation is needed.
Good faith	References that cooperation should operate in good faith, negotiation in good faith, etc.
Reciprocity	Mentions reciprocity in any facet of the cooperative process.
Notification	Mentions the need for notification between parties, most often in reference to prior notification.
Infrastructure	References that cooperation is over the development, construction, and operation of infrastructure.
Time	References some aspect of time, most often the length of time needed to build trust or negotiation an agreement.
Enforcement	Mentions the inclusion of enforcement mechanisms in the treaty or agreement.
Significant harm	Mentions the norm of no significant harm and its inclusion in an agreement or cooperative process.
Voluntary	References that a cooperative process, negotiation, etc. should be voluntary.

(*continued*)

Table A.1 Cont.

Key elements of cooperation	*Codebook description*
Uncertainty	Mentions that cooperation should address uncertainty or it provides examples.
International Court of Justice	Mentions the use of the International Court of Justice as a dispute resolution mechanism.
Compensation	References the use of compensation in exchange for allocation, harm, etc. in the cooperative process or negotiation.

Reference

Sadoff, Claudia W., and David Grey. 2005. "Cooperation on International Rivers." *Water International* 30 (4): 420– 27. https://doi.org/10.1080/02508060508691886.

Appendix B: Aspects of effectiveness

The aspects of effectiveness in Table A.2 that were identified from the qualitative content analysis are listed in the order of most commonly referenced using both the literature sample and the interview responses to the question defining effective transboundary water cooperation. The description provides a brief summary of the text that is captured in each aspect. Text can be coded under multiple aspects if it captures more than one idea.

Table A.2 Aspects of effectiveness

Aspects of effectiveness	*Description*
Achieve outcomes	Text advocates for measurable benefits, outcomes, results etc. to be achieved by the cooperative process to be considered effective.
Institutional capacity	To be effective, a cooperative process should have the institutional capacity in the form of RBO and/or agreement. Text mentions institutional capacity, RBO, or agreement.
Participation	Stakeholder participation and inclusion in the process are mentioned for effectiveness.
Define outcomes	An effective process defines what outcome needs to be achieved and they are agreed upon by the parties.
Mutual satisfaction	An effective process is when mutual satisfaction is held between the parties. This term also includes references to mutual benefits.
Context	Mentions that defining and evaluating effective cooperation is dependent on the context.
Implementation	Similar to "achieve outcomes" and "define outcomes," in that a cooperative process actually implements policy. It differs in that implementation does not necessarily include achieving the outcome.
Information and data	Mentions that effective cooperation as adequate data available, joint data collection, data exchange, trust in data and information, etc.

(*continued*)

Table A.2 Cont.

Aspects of effectiveness	*Description*
Procedure	Effective cooperation has procedural elements; generally, this refers to procedural rules in treaties, but could also include specific procedural rules within institutions.
Multi-scale	Effective cooperation is at multiple scales.
Reduce conflict	Cooperation is effective when it reduces conflict. This does not include text calling for dispute resolution mechanisms.
Time	Mentions that cooperation has adequate time to become effective, and the outcomes occur within a reasonable amount of time.
Shared interests	Mentions that it is effective if actors can develop or have shared or common issues, interests, goals, understandings, etc.
Adaptability	Effective cooperation is adaptable. Both in the outcomes determined by the parties and the process itself.
Power willingness	Mentions the role of political will and power difference in effective cooperation. It could be in any context with how these two factors can inhibit, encourage, or prevent effective cooperation.
Decision-making	Mentions cooperation is a process or is a decision-making process, or similar.
Future needs	Mentions that effective cooperation should consider, include, address, etc. future needs.
Dispute resolution	Mentions that effective cooperation has a mechanism for dispute resolution, which does not require that it has actually reduced conflict.
Integration	References text that addresses the integration of sectors, particularly with reference to IWRM.
Environmental change	Mentions that effective cooperation is that which achieves environmental change, or some other related environmental goal.
Technical cooperation	Effective cooperation includes cooperation at the technical level, may address technical cooperation more specifically.
Equitability	Includes references to effective cooperation as equitable, equal, having equity, etc. May also include specific references to the principle of equitable and reasonable use.
Sustainability	Includes references to sustainability and sustainable development and their variants, as a goal of cooperation or way to describe the process of cooperation.
Economic/political gain	Includes text that mentions economic or political gain as a reason to enter into cooperation, a potential benefit of cooperation, or why some may be more effective than others.

Table A.2 Cont.

Aspects of effectiveness	*Description*
Trust	Mentions that cooperation is effective if there is trust.
Shared model	Mentions that for cooperation to be effective there should be a shared and agreed upon model or conceptual model of the basin or aquifer. Most commonly stated with respect to a shared model for aquifers.
Negotiation	Includes text that mentions negotiation or re-negotiation with respect to effective cooperation.
Substantive rules	References substantive rules as needed for effective cooperation, either the term itself or an accepted principle that is considered a substantive rule.
Reciprocity	Effective cooperation is done with reciprocity.
Perception	Mentions perception of the cooperative process, perception of actors, of actors' needs or interests and the role perception plays in influencing effective cooperation or the perception of effective cooperation.
Scope	Effective cooperation should have a defined scope, including what issues are being addressed, or that the scope of many agreements should be broadened.
Regional pressure	Cooperation is more effective when there is regional pressure to cooperation, or regional encouragement, support, etc. e.g., the EU Water Framework Directive.
Compliance	Mentions compliance with agreement and other agreed policies.
Financial support	Includes all text related to funding or financing a cooperative process. Effective cooperation is generally seen as having adequate funding.
Responsive	Mentions that for cooperation to be effective the process and actors need to be responsive.
Human behavior	Effective cooperation elicits changes in human behavior. Such as the outcome stage in the Underdal 2002 figure.
Efficient	References how an effective process should be efficient, in the process and achieving the outcomes.
Accountable	References that the process and the actors should be accountable.
Lack of injury to people and system	Effective cooperation is cooperation that has a lack of injury to both the actors and to the water resource.
Consensus	Effective cooperation is achieved through consensus of the actors.
Interdependency	Mentions idea the cooperation is more effective if there is greater interdependence between the actors.
Transparent	Mentions transparency is needed for cooperation to be effective.

(continued)

Table A.2 Cont.

Aspects of effectiveness	*Description*
Legal	References legal effectiveness or legal tools in a more general sense.
Stability	Mentions that effective cooperation is stable or provides stability.
Pragmatic	Mentions that effective cooperation is pragmatic.

Appendix C: Barriers to effective cooperation

The barriers to effective cooperation in Table A.3 that were identified from the qualitative content analysis are listed in the order of most commonly referenced in the interview responses to the question asking for barriers to effective cooperation. The description provides a brief summary of the text that is captured in each barrier. Text can be coded under multiple aspects if it captures more than one idea.

Table A.3 Barrier to effective cooperation

Barriers to effective cooperation	*Description*
Context	Includes text that identifies human and physical contextual factors as barriers to effective cooperation.
Political will	Mentions how the political will of state actors can inhibit the development of effective cooperation.
National level	Mentions impact of national positions and national level policy, law, and efforts towards cooperation or lack of.
Donor financing	References impact of donor financing, and sustainable financing in general on effective cooperation.
Lack of knowledge	Lack of data on the water resource, including lack of data collection.
Trust	Lack of trust in the cooperative process, between actors, in data, etc.
Stakeholder	Not all stakeholders are included in the process or not included equally in the process.
Competing perceptions	Different perspectives on the resources, process, needs, etc. can inhibit effective cooperation.
Capacity	The capacity of actors can create unequal interactions. Text references negotiation, technical, human capacity, etc.
Power	Differences in power, hegemony, etc. influence effectiveness.
Sovereignty	State's tendency to prioritize sovereignty and national interests over shared interests, which can be a barrier to effective cooperation.

(*continued*)

Table A.3 Cont.

Barriers to effective cooperation	*Description*
Communication	References the lack of communication, not open or trustworthy communication, communication only over insignificant matters, etc.
Lack of resources	Mentions the lack of resources, including monitoring equipment, technical or scientific resources, etc.
Poor relationships	References poor, untrusting, etc., relationships between actors as inhibiting effective cooperation.
Sectoral scale	Mentions the challenge of dealing with many sectors related to water, e.g., water, disaster, agriculture, energy, etc. Also, might mention that the sectors could be at different geographic and governance scales.
Geography	In line with context, geography can inhibit effective cooperation, e.g., upstream/downstream, etc.
Rigid	References comments on the inflexibility of actors or the process as a barrier to effective cooperation.
Existing agreements	Mentions the potential for existing agreements to act as a barrier to effective cooperation.
Climate change	References the challenges posed by climate change as a barrier to effective cooperation.
Lack of institutional capacity	Mentions that the lack of an agreement, river basin organization, institution, or combination can be a barrier to effective cooperation.
Legal	Mentions lack of legal arrangements, or generally stated legal principles, can inhibit effective cooperation.
Process	Mentions if the process is not effective, other aspects will not be either, acting as a barrier to effective cooperation.
Unknown benefit	References that not knowing the potential benefits of cooperation will discourage cooperation from occurring or being effective.
Informed decisions	Having data and communication, and trust, to make informed decisions during the process.
Time	Without investment, engagement, or commitment in time, it is difficult for cooperation to be effective.
Compliance	Mentions compliance with agreements, without which is a barrier to effectiveness.
External actor interests	Mentions influence and potential competing interests of external actors to the cooperative process as a potential barrier to effective cooperation.
Openness	Lack of transparency and openness in the cooperative process can be a barrier to effective cooperation.
Potential for cooperation	Mentions the combination of timing, context, etc. that alters the potential for cooperation to happen.

Index

Note: Locators in *italics* refer to figures and those in **bold** to tables.

For Product Safety Concerns and Information please contact our EU representative GPSR@taylorandfrancis.com
Taylor & Francis Verlag GmbH, Kaufingerstraße 24, 80331 München, Germany

www.ingramcontent.com/pod-product-compliance
Ingram Content Group UK Ltd.
Pitfield, Milton Keynes, MK11 3LW, UK
UKHW020143200926
475702UK00005B/886

* 9 7 8 0 3 6 7 6 4 7 8 0 3 *